Whales and Seals
Biology and Ecology

Pierre-Henry Fontaine

4880 Lower Valley Road   Atglen, Pennsylvania   19310

Published by Schiffer Publishing Ltd.
4880 Lower Valley Road
Atglen, PA 19310
Phone: (610) 593-1777; Fax: (610) 593-2002
E-mail: Info@schifferbooks.com

For the largest selection of fine reference books on this and related subjects, please visit our web site at **www.schifferbooks.com**
We are always looking for people to write books on new and related subjects. If you have an idea for a book please contact us at the above address.

This book may be purchased from the publisher.
Include $3.95 for shipping.
Please try your bookstore first.
You may write for a free catalog.

In Europe, Schiffer books are distributed by
Bushwood Books
6 Marksbury Ave.
Kew Gardens
Surrey TW9 4JF England
Phone: 44 (0) 20 8392-8585; Fax: 44 (0) 20 8392-9876
E-mail: info@bushwoodbooks.co.uk
Website: www.bushwoodbooks.co.uk
Free postage in the U.K., Europe; air mail at cost.

Copyright © 2007 by Pierre-Henry Fontaine
Library of Congress Control Number: 2007925784

All rights reserved. No part of this work may be reproduced or used in any form or by any means—graphic, electronic, or mechanical, including photocopying or information storage and retrieval systems—without written permission from the publisher.
The scanning, uploading and distribution of this book or any part thereof via the Internet or via any other means without the permission of the publisher is illegal and punishable by law. Please purchase only authorized editions and do not participate in or encourage the electronic piracy of copyrighted materials.
"Schiffer," "Schiffer Publishing Ltd. & Design," and the "Design of pen and ink well" are registered trademarks of Schiffer Publishing Ltd.

Designed by Mark David Bowyer
Type set in Goudy OldStyle BT / Caslon224 Bk BT

ISBN: 978-0-7643-2791-9
Printed in China

Acknowledgments

I would like to express my thanks to Ursula Tscherter (Ores)who was kind enough to give me beautiful pictures of Minke whales. My thanks go too to Fabrice Kerleau and Gérard Mauger from Groupe d'étuddes des cétacés du Cotentin (GECC) for their picture of Bottlenose dolphins in action and of the necropsy of a Minke whale, to the members of the Réseau d'observation des mammifères marins(ROMM) of Rivière du Loup who gave me the opportunity to make necropsies of several seals and add to my documents many pictures of their anatomy, and with whom I spent several wonderful hours during the necropsies, to all the people from Fisheries and Oceans who gave me their steady help, and to the GREMM people for some beautiful necropsies and a financial help that was very much appreciated.

I would like to express my thanks, too, to the Fondation de la Faune du Québec, and its president, Bernard Beaudin, for their generous help. Thanks too to the Corporation des Pilotes du Saint-Laurent whose members play a very important part in assuring the safe navigation of the many ships going up and down a River where it is peculiarly difficult. They have a constant preoccupation in protecting a fragile environment and the species which live in it. They have contributed substantially to maintain this book at a reasonable price in spite of the many pages added.

Last but not least, a kind thought to my spouse, my children and all the friends who helped me in circumstances that would have made less strong-hearted people run away. And thanks to the editor, and his team without whose help nothing could have been done and who have been very patient with my last minute added or modified text and pictures.

Preface

It is hard to believe that almost ten years have passed since the first edition of this book was published, in 1988. To write that work, Pierre-Henry Fontaine drew on his vast personal knowledge of whale biology. His understanding of their ecology and their behavior at the surface of their watery domain was exceptional. As an accomplished scuba diver and skilled instructor, he had gained an intimate knowledge of their underwater world. Furthermore, Pierre-Henry was fully familiar with the documentation, ancient and modern, that covered all aspects of the life of whales and their relations with human beings. All in all, the first edition showed that his knowledge of, and experience in, cetacean morphology and anatomy was without equal. I can think of no-one else who has dissected and mounted so many skeletons of so many different species.

Over the last nine years, he has continued to study these leviathans and to stay abreast of advancements in the field of cetology (the study of whales). He has continued to dissect whales and to study their skeletal anatomy, be they the small harbour porpoises, the exceptional beaked whales, or the great rorquals. His passion for these animals, already evident and contagious fifteen years ago, has not only confirmed itself over the years but has intensified. And his enthusiasm is evident in this book as well.

Pierre-Henry marvelously illustrates the best of what the words "teacher" and "popularizer" can mean. He devotes as much energy and creativity to teaching as to learning. One of the features that distinguishes us the most from other animals is our ability to transmit cultural information. This ability is almost nonexistent in the other species, except for the few "cultural" transmissions observed in a rare cases. As far as human beings are concerned, cultural evolution predominates over most other forms of evolution. We are living today with the same body and the same mind that our ancestors, the Cro-Magnons, had 50,000 years ago. But because of language, writing, and teaching, our children will belong to the twenty-first century, not the world of the Cro-Magnons, Pierre-Henry Fontaine epitomizes what we must do to continue to evolve, rather than stagnate or worse, regress.

If authors such as Pierre-Henry were to restrict themselves to transmitting their knowledge, they would perform an invaluable enough service to society, especially if they did so with the kind of dedication, thoroughness, and clarity that characterize his work. But a book like this does more than transmit knowledge. To paraphrase the French geneticist Albert Jacquard's ideas on education, Pierre-Henry does not stop at assuaging our hunger; he makes us hungry to learn more. He informs and motivates not only experts but general readers as well. Popularizing cetacean biology involves an added responsibility since most readers will have to take so much on faith. And they can. The information presented here is on the cutting edge of what is currently known about whales. And the answers to many questions would not be possible were it not for Pierre-Henry's unique experience in cetology. This second edition provides detailed descriptions and illustrations not available anywhere else. The way that he popularizies cetology is in essence what science ought to be. In this book, you will find a thorough investigation that includes original information, a synthesis of what is currently known, a critical analysis of the prevalent theories, and an honest presentation of the grey areas requiring additional study.

This book contains a wealth of information presented in an accessible manner—a real joy to read. It will provide you with the pleasures of discovering as well as tapping into the mysteries that surround these animals, mysteries that Pierre-Henry Fontaine unrelentingly continues to shed light on through his work and why not, through the careers of naturalists or scientists inspired by this book.

    Cyrille Barrette
    Professor, Department of Biology
    Université Laval

Foreword

This book is an expression of my love for whales and more, for our world which could be so wonderful…as wonderful as those marine mammals. It is the product of observations, fieldwork, the gathering of specimens, and the compilation of information taken from the works of eminent cetologists, such as D.E. Gaskin (*The Ecology of Whales and Dolphins*), N. Bonner (*Whales of the World*), and Lyall Watson (*Sea Guide to the Whales of the World*), just to name a few.

It has been written to share with the reader, who has neither the desire nor often the time to invest in exploring libraries, the awe I experienced when I began to learn about these amazing animals so extremely well-adapted to their world, by giving him, as much as possible, the most accurate information, illustrated with great detail, and presented in a practical format.

This book is meant to instill in the reader the same fervent attitude I have for life in general, and whales in particular: each animal species is so precious that we must do all we can to ensure its survival.

We do not bequeath the Earth to our children; we have only borrowed it from them. This recently-heard comment has had a notable impact on me. Our knowledge of the biology and ecology of the animals that surround us can only help us to better understand them and to better protect the living heritage they represent.

If we want to give our children a world where life is worth living, let us learn and then take action: our knowledge can eliminate fear and free us to truly love.

I hope that this book will encourage readers to visit the most beautiful and peaceful animals on Earth. Once you have heard their powerful or seen, close-up, a 60- or an 80-ton whale perform a long, slow roll, so gracefully that the waters seem to have parted to let it pass, then the words *beauty, calm, power*, and especially *peace* take on their full meaning for you.

Introduction to the English Second Edition

ILE VERTE, SEPTEMBER 1997

Almost ten years have gone by since I first wrote this introduction, ten years to gather new documents, to collect and examine new specimens, and to continue acquiring knowledge on cetaceans. But with all this new wealth of information, my introduction could have very well remained unchanged.

Nonetheless, it seems appropriate and important for me to introduce this new edition. But why write a new, completely reviewed, corrected, and expanded edition, and for the first time ever, available in English?

Obviously, to correct errors that had slipped into the first edition, to include new findings and recent trends in the field of cetology, to take on subjects that had been initially neglected or omitted, but most of all, for my own personal satisfaction. Egotistical? Perhaps. But bear in mind that nothing provides me with greater pleasure than having the opportunity to share my knowledge, to allow neophytes from all walks of life to discover the wonders this science has to offer, by presenting them all the documentation I have gathered over the years.

This book pays homage to a professor of natural sciences who helped me become the man I have been over the last thirty years: a profoundly happy biology teacher. I cannot recall his name, but I have never forgotten his classroom, filled with skeletons, skulls, and all sorts of specimens that were always available for study. I remember his enthusiasm, his readiness to help and, more concretely, the first skeleton I mounted under his watchful eye. Undoubtedly, I have tried all my life to imitate him, in my collecting a myriad of objects to accompany my lectures. Some people may very well be appalled by some of the pictures and illustrations found within this book. They were not inserted to shock but rather to share some of my most intense moments, those that filled me with a joy hard to describe, such as seeing for the first time and handling (and smelling!) structures that, until then, I had only seen in illustrations or read about in books. These pictures are offered to you just as my teacher offered me my first skeletons, that is, to encourage you to find out more.

I have tried to give a maximum of information in the simplest format possible. I am certain that the general public is more interested in the workings of the animal machine than some "scholars" would have us believe, and that scientific popularization is not a brief overview of selected subjects but rather a clear, simple way to explain them. That is exactly what this book is about.

I could have devoted a good many pages to describing how we might feel when a sixty-ton fin whale, moving at fifteen kilometres per hour, surfaces three metres from our small boat and splashes us with its spout, or how we might feel when a beluga approaches, fixing us with its small, sombre eyes. Others can do that much better. Instead, I have chosen to take you on a journey into the world of the biologist that I am, into a world where we learn as much as we can about living creatures, where we learn to respect them without mythicizing, where, freed from fear by our knowledge, we learn truly how to love them.

If this book one day motivates any readers to study the natural sciences or if discovering the fantastic adaptations of cetacean anatomy and physiology, acquired over millions of years of evolution, proves to be a source of tremendous pleasure for others as it was for me, then I will certainly have achieved my goals.

In the twilight of my life, I realize that one of the greatest joys is the acquisition and the sharing of knowledge, and that it is always possible to discover and learn no matter how old we are.

With that in mind, welcome to the watery world of cetaceans…

Introduction to the English Third Edition

Why another new edition? I asked myself this question when I wrote the second version of my book Whales of the North Atlantic. I think it comes from the same desire to better a product which is, really, never finished. As time passes, one takes a step back and looks with a fresh eye on what was written, new documents become available, research makes progress. It is difficult to resist the temptation to share this new knowledge and not rewrite whole or in part, a book which, when written felt the best one could do.

I had been asked, many times, to include pinnipeds in my book. So, as time passed, I gathered plenty of documents which, at the time the second edition was sold out and had to be reprinted, were sufficient in quantity and quality to push me to completely overhaul Whales of the North Atlantic, Biology and ecology, and make of it Whales and Seals, Biology and Ecology.

I have kept a good portion of the text and illustrations from the second edition (1988) which were still adequate, and I have added information that was not available to me at the time, corrected some parts, and finally removed those which were out of date. Naturally I added, in different chapters, what was pertinent to the anatomy physiology and ecology of the pinnipeds, fur seals as well as true seals. I also added facts sheets for some seal species.

As in the previous edition I did my best to give the most scientifically correct information, using a language that will allow the non-specialist to have access to the information all the while satisfying the biologists wanting to use it.

Contents

Specific Environmental Adaptations of Cetaceans 13
    Density 13
    Viscosity 15
        Skeleton 16
        General Body Shape 28
        Body Surface 29
    Thermal Conductivity 31
    Swimming 36
    Pinnipeds Locomotion 41
        Walking on the Ground or Ice 41
        Underwater Locomotion 42
    Diving 42
    Marine Mammals and Nitrogen 50
        Toxicity 51
        Tissue Supersaturation 51
        Diving Adaptations in the Sperm whale 54
    Circulatory System 58

Perceiving the Environment 61
    Touch 61
        Cetaceans 61
        Pinnipeds 64
    Olfaction 66
    Taste 68
    Vision 69
    Hearing 73

Communication and Echolocation (Audiodetection) 81

Strandings 95

Nutrition 107
    Odontocetes 107
    Mysticetes 117
    Pinnipeds 139
    Marine Ecosystems and Food Resources 141

Migrations 147

Osmoregulation 155

Reproduction 161
    Male Genitalia 162
    Female Genitalia 168
    Mating 170
    Gestation 172
        Pinnipeds 175
    Lactation 176
    Growth 180
    Caring for the Young 181
        Pinnipeds 182

Parasites and Enemies 183

Whales and Humans 193
    Hunting or "Fishing" 193
    The Basques and Whale Hunting 195
    After the Basques 204
    Pollution 210
    Accidental Catches 212
    Whales as Resources 213
    Pinnipeds 221
    Research 225

Historical Photography 234

Paleontology 245
   Cetaceans Phylogeny 245
   Problems with Fossils 246
   Cetaceans Characteristics 248
      Early Archeocetes 252
      Adaptations to an Aquatic Life 253
      Development of the Tail as a Propelling Tool 255
      Remingtonocetids 255
      Kutchicetus, Remintonocetidae 256
      Basilosaurids 256
      Odontocetes 260
      Mysticetes 260
      Archaic Odontocetes 261
   Mysticetes 262
      Archaic Mysticetes 262
      Modern Mysticetes 264

Mounting Marine Mammals Skeletons 271
   Mounting 279
      Errors Made in Mounting Cetaceans Skeletons 280
      Causes of the Errors 281
   Techniques 287

Fact Sheets 291
   The Cetacea 292
   The Mysticetes 293
      The Bowhead Whale or Greenland Right Whale 294
      The Black Right Whale 296
      The Blue Whale 301
      The Fin Whale 307
      The Minke Whale 314
      The Humpback Whale 321
      The Sei Whale 328
      The Tropical or Bryde's Whale 331
   The Odontocetes 335
      The Beluga Whale or White Whale 336

The Narwhal	342
The Sperm Whale	346
The Pygmy Sperm Whale	352
The Northern Bottlenose Whale	355
The Harbour Porpoise	361
The Pilot Whale or Long-Finned Pilot Whale	365
The Short-Finned Pilot Whale	370
The Killer Whale	371
The White-Sided Dolphin	375
The White-Beaked Dolphin	378
The Common Dolphin	380
The Striped Dolphin	383
The Bottlenose Dolphin	385
The Pinnipeds	390
Harbour Seal	391
Harp Seal	396
Grey Seal	401
Hooded Seal	405
Walrus	409
Northern Elephant Seal	414
Glossary	423
Bibliography	434

SPECIFIC ENVIRONMENTAL ADAPTATIONS OF MARINE MAMMALS

The watery world in which marine mammals have evolved and thrived is far different from the terrestrial habitat of land mammals. Water is in fact almost a thousand times denser than air, about fifty times more viscous, and can transfer heat approximately twenty-five times faster. Also, because water is only slightly compressible, it is highly elastic: sound travels close to five times faster in water than in air.

DENSITY

One of the most striking features of whales is their size. Indeed, some species are the largest and heaviest animals ever to have walked the earth or swum the seas.

What are the factors that have allowed certain species of whales to attain such enormous sizes?

A terrestrial animal is supported by limbs that are for the most part directly connected to the vertebral column. This structure must be capable of bearing the animal's mass, which is subject to the force of gravity whether it is moving or stationary.

The overall size of a terrestrial animal is limited by the maximum size its limbs can attain and still remain functional. It is important to remember here that the mass of an animal increases as a function of its volume, which is the cube of its linear dimensions. On the other hand, bone resistance and limb muscle strength are surface functions and increase according to the square of their linear dimensions. Therefore, as an animal grows, its mass increases more rapidly than its skeletal strength. It is for this reason that an ant is proportionately much stronger than an elephant.

Drawing of *Indricotherium* and elephant skeletons (not to scale).

Diplodocus. British Museum.

It is impossible, then, for a mouse to become as large as an elephant. Without drastic changes to its body shape, its limbs would not be able to support a huge change in size; its mass would have increased at a rate much too fast for the corresponding increase in bone strength.

The limbs of a large terrestrial animal are proportionately much larger than those of a smaller specimen. They show almost no angulation between their segments; instead, they are arranged in vertical columns. This structure supports the animal's great weight and allows for a reduction in the mass of each limb segment. But all these structural adaptations are achieved to the detriment of the speed of limb movements and consequently, the speed of the animal itself. The loss of speed is compensated, at least in part, by the animal's size and stride length. The GRAVIPORTAL limbs of the elephant provide a good example of this structure. The largest terrestrial animals ever to have lived on earth, INDRICOTHERIUM (or BALUCHITHERIUM, a land mammal that stood more than five metres tall at the shoulder and possibly weighed more than 20 tons), and BRACHIOSAURUS (at around 78 tons, one of the largest of the dinosaurs) must also have had this type of skeletal arrangement. It is not likely that these gigantic creatures set any speed records.

Conversely, small or moderate-sized animals adapted for speed show a greater angulation between limb segments (i.e., a Z-shaped limb design), as a wolf skeleton illustrates. This type of skeletal architecture allows for greater and more rapid fore-and-aft movement of the limbs and thus more speed. In this case, however, the muscles support most of the animal's weight.

The bodies of marine mammals have a density that is only slightly higher, and sometimes even lower, than water. Thus, according to the Archimedean principle, a marine mammal is practically weightless in the sea, regardless of its size. We could easily imagine a diver standing on the ocean floor and lifting a whale in his arms! Unlike land mammals, which need support against gravity, marine mammals are, theoretically, subject to no limits on body size except, obviously, practical ones. As a

Elephants of the sea.

result, both small and colossal sizes can be found. For example, a Blue whale (*Balaenoptera musculus*), captured in 1947 by the Russian ship *Slava*, measured 27.6 m (90.6 ft) and weighed an estimated 197 tons (N. Bonner, 1980); by contrast, a Harbour porpoise (*Phocoena phocoena*) reaches a maximum length of 1.80 m (5.9 ft) and weighs at most 90 kg (198 lbs.)! It takes 2,000 porpoises to match the weight of one 197-ton Blue whale.

If some whales have remained small despite the advantages attributed to great size, it is probably because they feed on fast moving prey. In such cases, a great size would be more of a disadvantage. Furthermore, this wide range of sizes has allowed whales to occupy a vast array of ecological niches. In short, the conditions of microgravity in the aquatic environment allow for but do not necessarily call for large sizes. Factors other than the one mentioned above are involved and help explain why it is not surprising to find aquatic species of all sizes. In fact, pinnipeds, which are marine mammals, too, cannot attain the gigantic size of a Blue whale because they have to spend some time on land or pack ice, so gravity affects them. The largest of the pinnipeds are the male Elephant seals (*Mirounga sp.*) which do not exceed 4 tons.

VISCOSITY

As we noted earlier, water is approximately fifty times more viscous than air. This higher viscosity provides a much greater resistance to movement. We have only to try to run in water to convince ourselves of that! Such high viscosity does, however, provide the necessary support for the propulsory organs of aquatic animals, including the caudal fin of fish or cetaceans and the webbed feet of birds and pinnipeds.

How has the high viscosity of water affected the morphology and anatomy of marine mammals? We will have to make a distinction between pinnipeds and cetaceans.

SKELETON

Cetaceans

A whale's skeleton, in comparison to that of a terrestrial mammal, is strikingly simple.

Limbs and Girdles

The first difference we notice is in the limb structure. The forelimbs and hind limbs of most terrestrial quadrupeds are generally equal in importance. A few exceptions do exist, however, as the kangaroo demonstrates. But, in whales, this shared importance is clearly not the case: their forelimbs are greatly reduced in size and their hind limbs have, in most cases, completely disappeared. All that remains of the hind limbs is a vestigial pelvic girdle (pelvis or hipbones) which, no longer connected to the vertebral column, serves as a reminder of their former existence. This appendicular vestige now acts, at least in males, as a fulcrum, or support, for the genitalia.

Remains of the hind limb: the vestigial pelvis of a Blue whale.
Notice its position at the base of the penis.

A closer look at the forelimbs reveals the presence of only one movable ARTICULATION at the shoulder joint, between the SCAPULA and the HUMERUS. The elbow and the wrist joints no longer permit movement; they have become ankylosed, or fused together.

Cetacean hand bones are remarkably scattered. The five CARPI and PHALANGES are no longer connected. These bones are wrapped in a fibrous, rigid and resistant tissue. Although they are usually indiscernible, their outline is just visible in a few species. An example is that of the phalangic bumps of the Humpback whale. Odontocetes (toothed whales) still

Relics of the pelvic girdle of various cetaceans. The shape can be used to determine the species, sex, and age group (juveniles, adults, and old individuals).
1. { Harbour porpoise
2. White-beaked dolphin ♂
3. White-beaked dolphin ♀
4. Two Sperm whales
5. Northern Bottlenose whale
6. Beluga ♀ (3 years old)
7. Beluga ♀ (older adult)
8. Beluga ♂ (older adult)
9. Fin whale ♀
10. Fin whale ♂
11. Two Minke whales ♀
12. Blue whale ♀
13. Blue whale ♂ (juvenile)
14. Blue whale ♂ (old)

have five fingers, as do the Balaenidae (Right whales) of the Mysticetes group (whalebone or baleen whales). On the other hand, the Balaenopteridae (rorquals) have only four. Some fingers are elongated and have more phalanges (POLYPHALANGIA) per digit than do other mammals whose maximum is three per digit.

The morphology, relative size, and position of the forelimb indicate that this fin is used for balancing and steering rather than for propulsion. The presence of all the limb bones of the mammalian quadruped in whales undeniably demonstrates that cetaceans have evolved from a terrestrial mammal.

Pinnipeds

The pinniped skeleton looks a lot more like a land mammal, which is not that surprising if we consider that most of the species spend quite a long time on land or ice.

Forelimb of a sperm whale. It has five fingers, as do all Odontocetes.

Forelimb of a fin whale. It has only four fingers (the third one has disappeared), as do all rorquals.

We have to make a distinction between phocids on one side and Walrus and otariids on the other.

Phocids

The fore limbs of phocids are not used for propulsion in water. They are used to stabilize the animal, to make shorter turns, maybe to initiate the diving movement, and finally to move on land or ice. All articulations are fully functional: shoulder, elbow and wrist joints are perfectly mobile. The fingers end in powerful claws which will be used to prevent the animals from slipping as they climb on rocks or ice.

The role of the claws is very important in phocids.

The hind limbs are only used for propulsion in the water. The feet are completely projected toward the rear, because an apophysis of the *astragalus* prevent the foot from coming perpendicular to the leg The small toe is as long as the big toe, and digits 2, 3, 4, are relatively shorter. All the toes can spread far from each other during the propulsion movement. The feet then become very efficient flippers due to the skin which covers to them. The tibia and fibula are welded together and quite strong, because they support the muscles moving the foot. The femur is very short. The iliac crest is laterally everted, presumably providing a good anchoring point for the strong muscles used in lateral undulation while swimming. The thigh is not visible externally, so a good hydrodynamic shape is maintained.

Rear limb of a pinniped.

Front limb of a pinniped.

Astragale of an otariid.

Astragale of an phocid.

Foot in "propulsion" position (on the left) and foot in "recuperation" position (on the right) in the Grey seal.

Otariids and Walrus

Otariids swim with their forelimbs, and also use them to walk on solid ground. The Walrus uses them only when swimming slowly. They are mainly used for walking on land.

The scapula is robust, with two spines, the second one in the supraspinous fossa. This particularity is related to the way they walk on the ground, as otariids and Walrus put a lot of weight on the forelimb as they walk.

In otariids, the radius can no longer rotate around the ulna. In otariids, and to a lesser extent in Walrus, the fingers have only vestigial claws, and the last phalanx is prolongated by a cartilage rod.

The hind limbs of the Walrus are used to swim like those of phocids, and their feet look a lot alike, but, as in otariids, they can rotate forward and support the weight of the body, because the shape of the astragalus does not prevent this movement.

In otariids, the hind limbs are used only occasionally to swim, but they play an important part in terrestrial locomotion because the foot can rotate towards the front and support the weight of the animal.

Otary. Notice that its weight rests on its four members.   J.-P. Sylvestre

The Head

Cetaceans

The nostrils have, through the course of evolution, migrated to the top of the head. This position is beneficial for all air-breathing, fast-swimming animals, including those with a very large and heavy head. It allows the animal to breathe quickly by breaking the surface of the water without having to lift its head. Moreover, the anatomy of the whale's head has been deeply altered as a result of this migration: the snout (rostrum) is elongated; the PREMAXILLARIES (or premaxillae) and the MAXILLARIES (or maxillae) now overlap other bones in the skull such as the FRONTALS and PARIETALS, and extend to the OCCIPITALS. This telescoping of the skull has led to a superpositioning of the cranial bones rather than the usual juxtapositioning and should strengthen sutures while allowing prolonged growth. Opinions vary regarding the possible cause of these morphological changes. Some people see in them the effect of water's high viscosity on the head of the animal. Others see them more as the effect of changes to the neck and thorax musculature, an equally efficient locomotory adaptation allowing whales to swim in a medium that offers great resistance. Last but not least, the shortening of the neck, which was in progress all through cetacean evolution, brought with it a gradual loss of head mobility, so it may have helped in the displacement of the nostrils towards the top of the head, and so the bones telescoping.

Odontocete head.

Mysticete head.

Pinnipeds

The pinniped's head is not very different from the head of terrestrial mammals. The nostrils are still at the end of the rostrum, because their neck was not reduced during their evolution, maintaining great mobility of the head. The main differences are found in the orbital region and the teeth. The orbits are very large, to accommodate big eyes (adaptation for vision in poorly lighted environment) and the frontal part of the skull is very narrow between the orbits. Inside the nostrils the ethmoturbinates are greatly folded. As a result, the surface of the nasal mucosa is very great. We will see that it plays an important role in maintaining constant the body heat and that it reduces considerably the loss of body fluids. The teeth are all similar (homodonty) at least for the molar and premolar. These teeth have pointed cusps, and are used mainly to grab prey. Their alternate disposition (quincunx) does not allow chewing any more. The Walrus, whose regime is essentially *conchyophagic* is an exception. It has retained crushing molars.

Differences between the skulls of phocids, otariids and Walrus.

Otariids:

- Presence of a supraorbital apophysis, forming a prominent flat shelf above the orbits
- The frontal bones wedge themselves between the nasal bones, making a W-shaped suture
- Flat tympanic bullae
- Wide angular apophysis, placed at the base of the mandible ascending ramus
- No thinning of the posterior part of the mandible, no process there

Identifying characteristics of the otariid skull.

Phocids:

- No supraorbital apophysis.
- The nasal bones wedge themselves deep between the frontal, making a V suture
- Inflated tympanic bullae.

- Angular apophysis smaller, placed higher on the ascending ramus of the mandible
- The posterior edge of the mandible is thinner and makes a rear-oriented apophysis.

Identifying characteristics of the phocid skull.

Walrus:
- Maxillaries greatly enlarged to accommodate the very large canines
- No supraorbital apophysis, but an anteorbital one made from the frontal and maxillary bones
- Fused mandibular symphysis. The mandible cannot be separated.
- Vaulted palate, mainly in the anterior part.

Identifying characteristics of the Walrus skull.

Wolf skeleton.

Seal skeleton.

Harbour porpoise skeleton.

Vertebral Column

Cetaceans

The vertebrae are numerous and relatively simple in design. Some parts are shortened (i.e., neck region) while others are enlarged (i.e., lumbar region). But overall, the vertebral column is quite rigid throughout most of the animal's length.

As in most mammals (except SIRENIANS and BRADYPODIDAE), all cetaceans have seven cervical vertebrae. But the neck bones of cetaceans are compressed, and sometimes even fused into a solid bone mass (Odontocetes and balaenidae). The cervical vertebrae of the Beluga or white whale (*D. leucas*) and of the Narwhal (*M. monoceros*) have retained a high degree of flexibility. As a result, these animals have longer and more limber necks. This may be due to the fact that they live in waters often choked with floating ice floes. In general, though, the practically non-existent neck of most cetaceans minimizes mobility and prevents head bobbing during rapid movement in the water. This characteristic allows them to conserve energy as they swim.

The thoracic vertebrae (9 to 18, depending on the species) support the ribs, most of which, except the first four or five, lack solid attachment, either to the sternum or to other ribs. As a result, they are more flexible. In the Mysticetes, for instance, only the first pair of ribs is attached to the sternum. In most Odontocetes (that is, those in which sternal ribs are present), the first four or five pairs have this type of structural architecture.

Cervical vertebrae of a Minke whale.

Thoracic cage of a Mysticete.

Thoracic cage of an Odontocete. Notice the sternal ribs.

Whales have a high number of lumbar vertebrae (5 to 19 depending on the species). In addition to being quite large, these vertebrae usually bear highly developed TRANSVERSE and SPINOUS projections, or PROCESSES. These processes serve as anchoring surfaces for the huge muscle mass involved in the movement of the horizontally flattened caudal fin, the only propulsory organ of cetaceans. As opposed to terrestrial mammals, the fused sacral vertebrae associated with the pelvic girdle and the hind limbs are absent in cetaceans.

Lumbar vertebrae of a Harbour porpoise.

Caudal vertebrae and chevron bones of a Harbour porpoise.

The caudal vertebrae are also numerous and well developed. At their beginning, they bear prominent processes that progressively diminish in size and finally disappear over the length of the structure. Y-shaped bones called chevron bones, inserted between the first 12 to 15 caudal vertebrae, form a ventral arch to prevent blood vessels from being crushed as the muscles contract to lower the flukes.

The vertebral column of most terrestrial TETRAPODS allows for a great variety of movements. For instance, a cat falling back first will perform a quick turnabout to land safely on its legs.

The four stages of a turnabout performed by a cat falling back first (onto pillows!)

Who has not been amazed by the incredible bending and stretching exploits of circus contortionists? In striking contrast, the vertebral column of cetaceans is much more rigid, although substantial flexibility is conserved in its caudal region. As a whale swims, strong pressure is exerted over the length of the vertebrae, especially in the thoracic and lumbar area. The contact surface of their centra (or VERTEBRAL BODIES) is therefore broad and flat. The INTERVERTEBRAL DISKS are rather thin. Articular bony extensions (zygapophyses) on the spinous processes strengthen the union between the vertebrae and restrict bending. From underneath, a large ligament holds the vertebrae together to prevent compression of the vertebral column. Movement is thus quite limited in the thoracic and lumbar regions.

The contact (or central) surface of the caudal vertebrae, however, is more convex and the intervertebral disks thicker. Where the vertebral column enters the flukes, there is a special vertebra, the "ball vertebra," which has an intermediate shape between the previous ones, in which the greater axis is vertical, and the next ones, in which the greater axis is horizontal. It is there, in fact, that the vertebral column gains enough mobility to allow for the complex movements needed to propel the animal in the water. Adaptations to specific patterns in locomotion can be seen in all living creatures, as in birds, for example. Birds have an exceedingly long and flexible neck, but the rest of their body, all the way down to the caudal region, acts as a fulcrum. This particular design enables the trunk region to withstand the vigorous flapping of the wings during flight, and provides a solid support for the hind legs during landings. Moreover, since birds' forelimbs are so highly specialized as wings, their head and bill now perform manipulative and prehensile functions needed for activities such as feeding and nest building.

The skeleton of a whale does not have to support its weight and is thus relatively light and fragile. Most bones are made up of an inner, spongy region described as cancellous bone, which is covered by an outer and much thinner layer of compact bone. Whales can store large quantities of fat in these highly porous bones. In fact, the skeleton can supply one-third of the oil extracted from a whale. Moreover, some fresh bones contain so much oil that they float in water. Because the outer layer of compact bone is so thin, the skeleton of a beached whale is rapidly destroyed by wave action. All in all, the skeleton represents only 17% of the weight of the animal (22 tons for a Blue whale!).

Relative weight of the skeletal components of cetaceans:

Head and ribs	45%
Vertebral column	45%
Limbs	10%

Pinnipeds

The pinnipeds vertebral column has been a lot less modified than the rachis of cetaceans because they use their limbs to move on land and in the water. The neck has kept a good size in spite of the aquatic part of their lives, and the fact that they would save a lot of energy by having a shorter neck when in the water, like cetaceans do. It is probably because they spend quite long periods on land or ice that a long neck and the mobility it gives to the head allows them to watch more efficiently for their predators and gives them a better chance at survival.

There are some differences between the rachis of pinnipeds.

Phocids

As phocids cannot bend their feet under their body when on the ground, they have to crawl on the ground, using more or less their front limbs. Their cervical vertebrae (7) do not possess peculiarly strong neural and transverse apophyses. Thoracic vertebrae (15), have relatively small neural apophyses. Lumbar vertebrae (5), are large and possess strong transverse apophyses (3 times as long as they are wide to give good anchoring point to the muscles flexing the trunk). Their *zygapophyses* are loosely articulated to allow lateral flexion of the body during swimming. As phocids use reciprocating movements of their feet in a horizontal plane to propel themselves in water, it induces lateral movements of the body. There are 3 vertebrae in the well developed sacrum, and the caudal vertebrae (10-12) are small.

Cervical comparison of the otariid (left) and the phocid (right)

Lumbar comparison of the phocid (left) and the otariid (right)

Otariids

Otariids move on land by using their 4 limbs. Their very close feet can bend forward and support part of their weight. They are, however, so near to each other, and their hind limbs so short, than when they walk on land they use mainly their front limbs, helped by oscillations of the neck, which has become very powerful. In water, they use mainly their forelimbs. Accordingly, their cervical vertebrae (7) are larger than those of phocids, with larger apophyses to give good anchor points to stronger muscles. The first thoracic vertebrae neural apophyses are a lot longer, too, to give the strong elevator muscles of the neck room enough to attach. There are 15 thoracic vertebrae. Lumbar vertebrae (5) are smaller than those of phocids. The transverse apophyses are smaller too (as long as they are wide). The zygapophyses are tightly articulated, limiting trunk movements. As the forelimbs are used to swim, the need for a horizontally flexible lumbar region does not exist. The caudal vertebrae (10-12) are small.

Walrus

The spinal column of the Walrus is somehow in between the two former examples. Cervical vertebrae are smaller than otariids but larger than phocids, as are the neural apophyses of the first thoracic vertebrae. There are 6 lumbar vertebrae with long transverse apophyses, for the Walrus swims with its feet and by undulating its body in the horizontal plane, as phocids do.

GENERAL BODY SHAPE

Cetaceans

To move rapidly and efficiently through water, cetaceans obviously need a HYDRODYNAMIC body shape. Animals will expend much less energy if they can glide through the water rather than have constantly to move forcibly through it. Whales have achieved this ease of movement through the development of an essentially cylindrical body, tapered at both ends. When we compare the body shapes of a dolphin, a shark, an ICHTHYOSAUR, and a swimming penguin, we cannot fail to notice that these animals share many morphological characteristics. These animals all belong to different taxonomic classes and present a striking example of convergent evolution. Since these organisms face similar problems, it is not surprising to find that their solutions, all products of natural selection, are also very much alike.

Shark, ichthyosaur (an extinct reptile), Harbour porpoise, and penguin: a good example of convergent evolution.

Anyone who has seen a whale will agree that its tail end is indeed tapered. But what about its head? Some head shapes, particularly in Odontocetes, do not go very well at all with the fusiform shape apparently ideal for movement in a viscous medium such as water.

Toothed whales bear on their forehead a lens-shaped body called the MELON. Its function is the subject of much debate. One hypothesis suggests it is an acoustic lens, another, a hydrostatic organ. Or could the melon's function be analogous to that of a ship's curved bow, important for reducing turbulence?

Pinnipeds

Although they spend part of their life on ground or ice, pinnipeds have a very good hydrodynamic shape, which is not far from the one exhibited by cetaceans.

BODY SURFACE

Cetaceans

In keeping with their fusiform body shape, cetaceans have eliminated all protrusions from the body surface, apart from the forelimbs. The EXTERNAL EARS, or pinnae, have disappeared. As for the auditory canal, it is highly reduced and appears as a pinhole on each side of the head.

The nipples are contained within mammary slits (or grooves) and protrude only in lactating females. The vagina and anus are hidden within a GENITOANAL SLIT.

Vibrissa at the tip of a Minke whale's mandible.

In males, the penis, visible only during erection, is completely retracted within the body. The testicles are internal and the scrotum has entirely disappeared.

Cetaceans have extremely soft and smooth skin. All cutaneous glands such as SEBACEOUS GLANDS are lacking. Hair has almost totally disappeared, although some can still be found on the head, the lips or under the chin of a few species.

In Odontocetes, only the hair follicles are found.

Normally, hair plays an important insulating role and provides protection against abrasion. Bald men are painfully aware of the vulnerability of their scalp to abrasive surfaces. But this type of protection is unnecessary for whales since they shun the shore. Furthermore, hair covering would considerably increase the animal's drag and thus its resistance to movement.

The high sensitivity of a whale's skin is probably due both to its lack of hair and its relative thinness. Whales will therefore avoid, as much as possible, potential sources of danger to their skin (rocks, boats, wharves, buoys, etc.). Exceptions do exist, however: some whales have indeed been seen rubbing themselves on stationary boats. Were they doing this to rid themselves of an itch caused by parasites such as the highly specialized crustacean we analogously call "whale-lice" (*Cyamus sp.*)?

The skin and blubber of a cetacean are firm without being rigid. The subcutaneous fat is not only made up of adipose tissue, as it is in pigs, for example; it also has numerous layers of intertwining fibres of connective tissue. These fibres apparently serve to store energy and significantly reduce muscular efforts during active swimming (*Discover*, March 1997). Many biologists believe that all these characteristics play a crucial role in determining how water flows around the body of a moving whale.

Head of a Beluga. The melon is very visible.

Beluga. Opening to the auditory canal.

Beluga. The genitoanal slit and those containing nipples.

Fin whale in the St. Lawrence River.

Pinnipeds

All pinnipeds have a skin covered with hair. The pelage of an animal is usually made of 2 kinds of hair: long and coarse guard hairs and soft more densely packed, undulated underfur hairs. Hair grows a certain amount of time, then falls off. Hair may be lost all year long or shed over a short period of time called molting. Pinnipeds molt once a year. Hair protects again heat loss, especially in otariids which have a relatively thin layer of fat. Otariids do not dive very deep either. In the case of phocids, hair may act more as a mechanical protection for the skin than as an insulating layer. They have more fat under the skin and their fur is less dense. They usually dive deeper, too.

THERMAL CONDUCTIVITY

Water transfers or disperses heat twenty-five times faster than air. Although we tend to complain about the heat when the air temperature is 25°C, water at that temperature seems cold to us. Any animal that must regulate its body temperature will face a much greater problem if it lives in water rather than on land.

Terrestrial animals have developed many ways to keep warm. Some bear a thick fur covering for protection, while others, rather than having to struggle against the cold, seek shelters and sleep the winter season away. A significant drop of body temperature may accompany this winter sleep, sometimes reaching just a few degrees above the surrounding air temperature. Animals that practise this type of dormancy or torpor, such as ground squirrels, chipmunks, and groundhogs, are said to hibernate.

Marine mammals face a different problem. They live in an environment where temperatures remain relatively more constant. In fact, temperature differences in all oceans are far less than those encountered on land. In spite of this, some whales will use these differences to the advantage of their life cycle, by benefiting from the abundance of food in the deeper, colder waters in the summertime and exploiting the warmer equatorial waters to give birth and nurture their calves during the winter months. Nonetheless, even in tropical waters, especially at lower depths, marine mammals, being warm-blooded animals that maintain a relatively constant body temperature (HOMEOTHERMS), must find ways to reduce heat loss.

Having a large size confers many distinct advantages. One that carries significant biological consequences, especially in larger whales, is having a smaller surface-to-volume ratio. We have already mentioned that the volume of a body increases more rapidly than its surface: the larger the animal is, the smaller its surface will be, in proportion. Animals tend to lose substantial heat through their skin, a loss that increases considerably if they live in a medium that transfers heat as well as water does. Therefore, the smaller the body surface exposed to water, in relation to the animal's volume, the smaller the energy loss by conduction. Moreover, the metabolic needs of a large animal are proportionately lower than those of a smaller animal, as the following numbers indicate.

Animal	Body Mass (kg)	Metabolic Rate (Basal metabolism)
Guinea pig	0.6	223
Rabbit	2.0	58
Man	70	33
Ox	600	15
Elephant	4,000	13
Blue whale	130,000	3.5

Note: The metabolic rate is expressed in kilocalories per kilogram of body mass per day.

The great rorquals consume 4% of their total weight per day in the summer. During this feeding period, which lasts about 120 days, they gather in the colder, food-laden waters. The remaining time is spent in the warmer and less productive waters. On an annual basis then, they consume an average of 1.5 to 2% of their body weight each day. These whales eat about five times their weight per year. By contrast, a human being eats the equivalent of 15 times his weight in a year. The pygmy shrew (a small insectivore), which is the smallest living mammal, with a weight of only a few grams, consumes a little more than the equivalent of its body mass each day!

Marine mammals, then, tend to be large animals in order to limit the surface exposed to heat exchange. Even then, a large size is not enough. They also have to insulate their organs, muscles, and viscera from any contact with water. There are many ways to accomplish this task. The most common defense mammals have against cold is to grow thick pelages, or fur coats, that are more or less covered by oily secretions which traps an insulating layer of air.

Polar bears, beavers, sea and river otters, fur seals, and true seals have adapted in this particular way. For cetaceans, which expend energy only in water by swimming rapidly and perform deep dives, there must be other avenues of heat conservation. For them, a thick pelage would only offer greater friction or resistance to movement and, in turn, would call for an increase in energy expenditure. Moreover, since air is compressible, its insulating power when trapped by a fur coat would rapidly decrease along with the increasing HYDROSTATIC PRESSURE at greater depths. (Hydrostatic pressure increases the equivalent of one atmospheric pressure every ten metres). Divers using a wet suit to trap air experience

The skin and blubber of this beluga have been removed. Notice that all of its adipose tissue is between the skin and the muscles.

this phenomenon during deep dives in cold waters. The fact that Pinnipeds have kept their fur in spite of the augmented drag is caused by the fact that they are regularly on land or ice.

To keep warm, whales use a strategy that is, by the way, also employed by their terrestrial cousins: the accumulation of a thick layer of fat (blubber) immediately beneath the dermis.

An emaciated or thin whale at Percé.

This layer is in fact part of the skin and makes up the hypodermis. It is always surprising to see the ease with which blubber is ripped from the underlying muscles of a dead whale (a process called flensing). The thickness of the blubber layer varies considerably among individuals of the same species and among different species as well. The Blue whale (*Balaenoptera musculus*) and the Sperm whale (*Physeter macrocephalus*) may have between 15 and 20 cm (6 and 8 in.) of blubber. On the other hand, the Right whale (*Balaena mysticetus*) may have more than 50 cm (20 in.). In fact, it is because of this characteristic that the latter were christened "right" whales (because they were the "right" whales to hunt). Since fat has a low density, these whales actually floated once dead, which naturally made salvaging carcasses much easier! Blubber in water has three-fourths the insulating power of air. Cetaceans are thus well protected in their natural environment. This layer of fat also serves as an important food reserve. In spring, when Humpback whales (*M. novaeangliae*) first appear in Canadian waters, they have less fat than when they migrate south the following fall. The reason is that in their winter breeding grounds, on the Silver Bank off the shores of the Dominican Republic, there is very little feeding. It should be noted, however, that the blubber, which acts as thermal insulator, will be used as the last extremity. A fasting whale will first use the fat stored in its bones (and we have seen that it represents 1/3 of the oil recuperated from a carcass). Then they will use their muscle mass as food. I have necropsied a Beluga whale very emaciated, it had lost maybe 1/2 of its muscle mass, but still had blubber 5 cm thick, which is about normal for a Beluga its size.

The percentage of lipids in the blubber can augment its isolating power and compensate in some measure for the effects of a smaller size. The Harbour porpoise (*Phocoena phocoena*), a small species living in temperate to cold waters has a lipid content in its blubber of approximately 81.6%. The Spotted dolphin (*Stenella attenuata*) a pantropical larger species has only around 54,8% of lipids in its blubber. The slightly thicker blubber of the Harbour porpoise has 4 times the insulating power of the one from the dolphin

All current studies suggest that the basal metabolism of whales and pinnipeds is equivalent to that found in other mammals, that is, 984 to 1,032 kcal/m^2 of body surface/24 hours.

The insulating power of the subcutaneous blubber layer of cetaceans is so great that the body temperature of a dead whale can momentarily rise enough to literally cook its flesh. As a result, the muscles will easily detach from the bones, when flensing and butchering are not performed immediately after the death of the animal. The heat from the metabolic activity at the moment of death and from subsequent putrefaction cannot escape to the outside because it is trapped by the blubber. This phenomenon is known as the Norwegian haybox principle. A dead whale will literally melt when exposed to the sun, releasing vast quantities of oil through the holes left by scavenging seabirds (or by biologists!).

I have noticed this phenomenon on various occasions on Île d'Anticosti (Québec, Canada). In late May 1985, while salvaging the skeleton of a stranded Blue whale in Anse aux Fraises, three weeks after it had beached, I noticed that its flesh was extremely hot to the touch. Upon the first incisions, large quantities of oil flowed from the whale's body and made walking on the flat reef, which encircles the island, perilous over a distance of two kilometres!

Similarly, in June 1992, it was easy to observe, from an airplane, great quantities of oil that had spilled from the carcass of a stranded Sperm whale, near cap Blanc on île d'Anticosti, Québec.

The insulating power of blubber is so efficient that it may create problems when whales increase their muscular activity by even a small degree of intensity. Cetaceans, like all mammals, are HOMEOTHERMS, and a significant rise in body temperature from increased activity can have harmful consequences such as heat-stroke or a debilitating fever, as we know from our own experience.

Sperm whale stranded near Port-Menier, île d'Anticosti. The ebb tide washes away the oil flowing out of the carcass. A vehicle is parked to the left of the animal.

Therefore, an efficient system is needed to enable a whale to maintain a constant temperature, no matter what it does or where it lives. It is the blood flowing through the ARTERIES, VEINS, and CAPILLARIES that acts as a cooling agent. By flowing close to the surface of the skin, the blood is cooled by the surrounding cold water and returns deep inside the body to cool the different organs.

At the same time, a system that prevents blood from cooling the body when whales have to conserve heat (during periods of inactivity in cold water, for example) must also exist. Blood flows through blubber by means of arteries that ramify into numerous cutaneous capillaries. These capillaries regroup to form veins that eventually return the cooled blood to the body core. If a whale has to conserve heat, its autonomic nervous system causes the arteries in the blubber to contract. A small amount of blood circulates in the skin to maintain cellular integrity, but most is redirected to alternate deeper veins. Consequently, the blood that returns to the body core has not been significantly cooled by the water and will not, in turn, cool the body.

An even more highly specialized system prevents heat loss through those organs that have little or no fat, such as the tail, dorsal fin, and forelimbs. Arteries that carry blood into a tissue ramify into a plethora of small capillaries that parallel and intersect a series of similar venous vessels forming a network called RETE MIRABILE. This vascular network allows warm arterial blood to transfer heat to the colder venous blood returning from the skin. Consequently, the blood going back to the body remains at core temperature, and heat loss to the outside is considerably reduced.

Diagram—Countercurrent Heat Exchanger

Arteries and venous plexus of a Blue whale, île d'Anticosti, July 1987.

Cross section of the tail of the Blue whale "Pita," showing an artery surrounded by its venous plexus. Île d'Anticosti, August 1992.

That system is the countercurrent heat exchanger principle, which was discovered by humans about 200 years ago, but invented by nature millions of years ago. When the animal has to cool down, blood flows into and dilates the arteries, with the resulting compression of the surrounding veins. The blood is thus redirected to normally compressed veins that carry the cooled blood directly to the general circulation. As a result, the animal can cool down.

SWIMMING

Cetaceans are exceptional swimmers, propelling themselves by the up and down movements of their tail (caudal fin); their forelimbs are used only for balancing and steering. The caudal fin consists of a horizontally flattened surface. At its center, we find a core made up of an extremely dense and fibrous tissue whose extensions are firmly attached to the caudal vertebrae. An envelope of ligaments that splay out from the vertebrae surrounds this fibrous core. These ligaments, essentially inelastic, form an incredibly solid structure—a whale weighing many tons can be suspended with a wire that passes through holes made in the caudal fin!

Siam, a Humpback whale, showing its caudal fin, off the shores of Baie-Sainte-Catherine, Québec.

The epaxial muscles attached to the spinous apophyses of the vertebrae, above the transverse apophyses, serve to lift the tail. They are much larger than the hypaxial muscles, which are on both sides of the column under the vertebrae and are responsible for pulling the tail down. The disparity between the two muscle masses could suggest that propulsion is mostly the result of the upward movement of the tail, but it is now really contested. Hypaxial muscles include most of the muscles which control the angle of attack of the flukes. It allows for the fine adjustments necessaries to meet all the situations encountered while swimming.

Tail of a sperm whale.   ORES–Guillame Pelletier

Whales are incredibly powerful animals. Sei whales (*Balaenoptera borealis*) may reach speeds of 64.7 km/h (40.2 mi./hr or 35 knots) over short distances. The great rorquals (*Balaenoptera musculus*, *B. physalus*) swim at 26 km/h (16 mi./hr or 14-15 knots) with top speeds reaching 37 km/h (23 mi./hr or 20 knots). The much smaller dolphins and porpoises can attain similar speeds, disproving Gray's paradox. After an experiment in which he towed dolphin models through the water, this cetologist concluded that in order for them to attain such speeds, their muscles would have to perform at ten times the level of those of terrestrial mammals.

It is in fact a pseudoparadox, seeing that Gray (1936) did not consider that water flows differently along the rigid model of a cetacean than along a living dolphin.

Fin whale close to Saguenay. There is little movement around the animal, though it is moving quickly.

When a fluid moves around either a stationary or a moving object, it is slowed by the friction of its particles striking the surface of the object.

The thin, film-like layers of water farther from the object's surface flow more rapidly than those closer to it. As long as this flow is not too fast, these layers remain parallel. This phenomenon is known as a laminar flow pattern. But if the object's speed increases, the layers separate and eddies are formed. A turbulent flow pattern is thus generated. The striking of water particles against the object's surface reduces the flow speed all the more as turbulence increases.

But the flow pattern around a living dolphin remains laminar, even as the dolphin's speed increases. The animal's drag (resistance to movement) is therefore considerably less than that of a dead dolphin or a plastic model being hauled at similar speeds. Its muscle power and energy expenditure are therefore hardly exceptional or even paradoxical!

How do cetaceans maintain a laminar flow around their body? Their streamlined body shape has much to do with the answer. But a more complete explanation lies within its skin structure. According to some biologists (Kramer, 1960 and 1965), the skin of cetaceans is made up of a smooth elastic epidermis that is attached to the underlying dermis by longitudinal ridges, similar to the ones responsible for our fingerprints. These ridges are associated with capillary bundles and water-filled microtubules in the epidermis. The surface of the skin can therefore respond instantaneously, almost simultaneously, to changes in pressure, and eliminate, one by one, the microwhirlpools created when these animals swim. This action on the skin assists the maintenance of a laminar flow and drag is substantially reduced.

In addition, skin cells are filled with oil droplets that lubricate the surface of the body and facilitate the flow of water, whose density is lowered by a slight temperature increase in the immediate vicinity of the skin. Much like the keel of a boat, the dorsal fin controls the backward flow of water as the animal dives beneath the surface. It should be noted that slow-moving whales, such as the Right whale, lack dorsal

Skin Structure of Dolphins

fins. Finally, the presence of microscopic grooves (whose design is not due to chance), the shape of the pectoral and caudal fins, and the ability of whales to change course to adjust to various flow speeds, complete the list of necessary adaptations for rapid movement that bring Gray's "paradox" down to size.

These hypotheses are, however, challenged by various cetologists, cited by Fish and Hui (1991) in a scientific journal updating the many years of research on cetacean locomotion. Still subject to controversy are such factors as: the laminar flow of water along the cetacean body, the role of the skin in the control of turbulence, the speeds that dolphins reach, and the belief that only the upward movement of the tail serves in propulsion. Fish and Hui acknowledge that research is far from complete and that much remains to be done before indisputable conclusions concerning swimming in cetaceans are drawn.

Regarding the propulsory function of the tail, studies of a film sequence of swimming Killer whales (*Orcinus orca*) show that the extremities of their flukes curve downwards. We might naturally assume that the downward movement is thus the driving force. But this motion has other uses too. It is a downward tail slap that Killer whales near Norway use to stun the herring on which they feed! When a Humpback whale tail-slaps the water (*lob-tailing*), it does so on its back. Is this done to increase the force with which it smacks the water?

Blue whale in the Grandes-Bergeronnes area.

Pierre-Michel Fontaine

Ann Pabst (*Discover*, March 1997) at the University of North Carolina has proposed a new explanation for the remarkable efficiency of cetacean locomotion. When dolphins are trained to exercise by swimming against a force sensor (dynamometer), their oxygen consumption increases with the effort given until a levelling-off effect is reached. This situation is in itself quite normal. It simply means that the animals have reached the limits of their aerobic metabolism. To generate more force, they have to resort to their anaerobic production of energy, a response typical of most living beings. But this production of energy without oxygen is short-lived because it normally also creates lactic acid (a by-product that interferes with muscular activity). Consequently, when dolphins are subjected to strenuous exercises, we should notice a rise in the level of lactic acid in their blood. Surprisingly though, blood samples reveal only low levels of lactic acid, even after doubling the pressure on the dynamometer. The explanation for this seems to lie in the function of dolphins' muscles, perhaps similar to that of a kangaroo. Once a kangaroo has reached its aerobic capacity of energy production, it can double its hopping speed without burning additional fuel in its muscles. But how can it accomplish such a metabolic feat? It does so by simply letting the tendons of its legs take over. As the animal touches the ground after a jump, the tendons of its legs are stretched, storing a certain amount of energy within them. Most of that energy (93%) can then be used for the next jump. In dolphins (and in most cetaceans), the muscles are wrapped in a sheath of crisscrossing collagen fibers. Part of the muscular fibres is attached to this sheath and the rest to the skeleton. The length and the angle of the fibers with respect to the muscle axis suggest that the sheath could act as a spring, accumulating energy at certain moments and expending it at other times. An almost parallel arrangement of

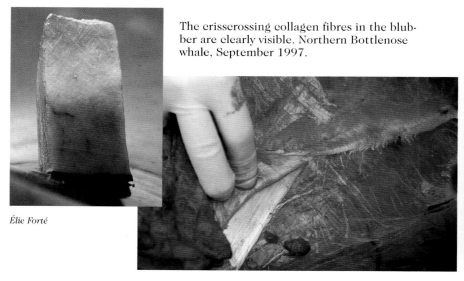

The crisscrossing collagen fibres in the blubber are clearly visible. Northern Bottlenose whale, September 1997.

Élie Forté

The crisscrossing collagen fibres envelop the epaxial muscles of a cetacean. Northern Bottlenose whale, September 1997.

the collagen fibers in the blubber layer may add to the efficiency of this mechanism. Although much more work is needed to validate this hypothesis, it, nonetheless, seems quite reasonable.

Young whales are not endowed with the same muscle power that adults have. Regardless of that fact, calves not only maintain the swimming speed necessary to keep up with the adults, but do so without really tiring. By swimming close to and below the mother's dorsal fin, which happens to be just behind the largest part of the whale's body, a calf can benefit from the "Bernoulli effect." In other words, the speed of the water flowing between the two bodies is increased and, as a consequence, the pressure in this area drops. The calf is therefore "drawn" toward its moving mother and saves more than 25% of the energy otherwise needed for it to keep up.

Similarly, to save energy, many cetaceans will ride pressure fields created by the wake of a ship or the head of larger whales. Depending on the shape of the wave created, whales will adjust the position of their flukes up or down, in order to be pushed forward using a minimum of energy, a little like a surfboarder riding a wave. In this type of movement, the role of the pectoral fins in maintaining stability is very important.

PINNIPEDS LOCOMOTION

Pinnipeds do not use their tail, which is very short to swim. They will use their limbs in a different way whether they are fur seals or seals, or they are swimming or walking on land.

WALKING ON THE GROUND OR ICE

Phocids

Seals, as we said before, cannot bend their feet under their belly. They are always directed toward the rear and cannot be used on the ground. Seals move on land by undulating their whole body, a little like a big caterpillar. The forelimbs, which possess big claws can be used as an anti-slip device. Their fur can also prevent them from slipping on the ice. There was a time when one would put seal skin under his skis to climb back the slope he had just glided down. It was before T-bars were invented, evidently.

Otariids

Fur seals, whose feet can be pulled under the body, will use them on land. The two feet, which are almost attached by the heels, are not very mobile, When a fur seal moves on land it projects its head and neck forward. The center of gravity is moved forward too, and the feet are then lifted from the ground and projected forward by flexing the trunk. It only has to lift its head and start again. Those movements can be done fast and allow for a good speed on the ground.

UNDERWATER LOCOMOTION

Phocids

When they move through water, phocids use mainly their hind limbs. They use their forelimbs only to steer or to help them to dive when they are on the surface. I have seen young Harp seals (*Pagophilus groenlandicus*) in aquariums, that were using them to move very slowly or to remain in the same spot. When they swim, they use their feet. Those are able to open widely so they can produce good thrust. Fingers I and V are of same length, fingers II, III, and IV are smaller, and all are united by a web of skin that makes a good fin of the foot. Feet can also fold tight during recuperation movements so they do not offer resistance to water. They are used alternatively, with a lateral flexion of the pelvis. The very hydrodynamic general shape of phocids reduces drag very efficiently.

Walrus

The Walrus move through water as phocids do, using mainly their feet and lateral undulation of the pelvis for propulsion. Similarly, they use their forelimbs only to swim slowly or to change direction. On the ground they use their limbs somewhat like otariids.

Otariids

Otariids swim with their forelimbs using them as birds use their wings to fly. Hind limbs are essential for changes of direction, but are not used for propulsion.

DIVING

Marine mammals swim for a variety of reasons: to move, to court, sometimes (it would seem) to "play," and of course, to feed. In order to feed, most marine mammals have to perform dives at various depths.

Fin whales blowing near the Saguenay River.

When they dive, they cannot use the oxygen in water for their metabolic needs, but must make do with the oxygen taken in at the surface. What are the adaptations that allow marine mammals to remain active during these long periods of APNEA?

A man with no particular training can hold his breath for approximately one and a half to two minutes (the record for static apnea is more than 7 minutes). The platypus can hold its breath for 10 minutes, the hippopotamus, 15 minutes, and the beaver, 20 minutes. But a rorqual has the ability to stay submerged for 40 minutes, the Sperm whale, more than 90 minutes, and at least one Northern Bottlenose whale (*Hyperoodon ampullatus*) managed to remain underwater, after having been harpooned, for a period of 120 minutes. We might easily assume that whales have extremely large lungs that enable them to store vast quantities of air. The 3,000-litre lung volume of the Blue whale would seem to justify this assumption. However, the lung-to-body weight ratio reveals quite another story. It shows that, proportionally, whales have, in fact, slightly smaller lungs than the average terrestrial mammal. Human lungs represent about 1.75% of the organism's total body mass, those of an elephant 2.55%, those of a sperm whale, 0.91%, and those of a Blue whale, only 0.73%. Seals are also able to do very long dives. In fact, considering their size, they are champions. A female southern Elephant seal (*Mirounga Leonina*) remained submerged for 120 minutes, a Weddell seal (*Leptonychotes weddelli*) for 82.

It is not by storing vast quantities of air in its lungs that a Sperm whale, for example, can stay submerged for 90 minutes. Since air is compressible, it would actually be dangerous for whales to carry down large volumes of air. Hydrostatic pressure increases by one atmosphere (1 kg/cm^2, 14 psi) for each ten-metre (33-ft) increment in depth. If a large volume of air were contained in the thorax, this air would compress more and more with the rising hydrostatic pressure. As a result, its volume in the lung would continuously decrease. The thoracic cage would undergo mechanical constraints difficult to bear. Having relatively small lungs is then more advantageous for marine mammals. Furthermore, contrary to terrestrial mammals, cetacean lungs are more dorsal (in their thoracic cage) and their diaphragm, the tendinous muscle that separates the lungs from the abdominal cavity, is also very oblique. The particular positioning of the lungs and diaphragm allows for a better balance in water and for the air-filled lungs to empty almost completely.

An engraving illustrating the common belief that water spouts from the blowholes of whales.

Blows of Fin whales.

Cetaceans can ventilate their lungs much more efficiently than terrestrial mammals. In fact, 85 to 90% of the air contained in the lungs is exchanged, as opposed to 10 to 15% in the normal breathing of terrestrial mammals. Another notable difference is the speed at which the air in its lungs is exhaled. In four seconds, a Fin whale exchanges 3,000 times more air than a human does! When a whale surfaces and exhales, a blow is seen. This exhalation occasionally begins just below the surface of the water and carries with it a large quantity of water. It is this maneuver which has created the misconception that water comes out of the blowholes of whales.

Nostrils of a Fin whale. They are open and the splashguard is clearly visible.

Water is re-directed toward the sides of the splashguard.

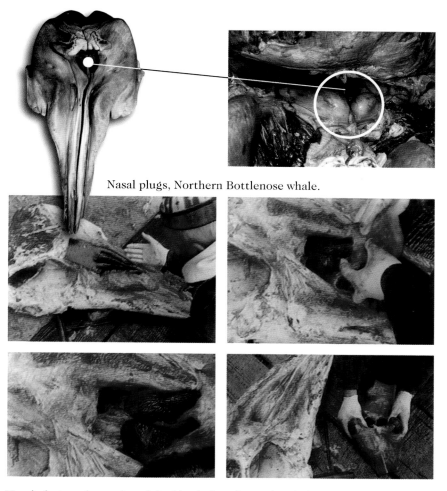

Nasal plugs, Northern Bottlenose whale.

Nasal plugs and muscles of the blowholes of a Minke whale.

Contrary to what it seems, the spout is made up of air saturated with water vapor that probably condenses from its violent expansion during exhalation (ADIABATIC RELEASE) and its contact with the colder atmospheric air. In addition, it most likely contains both an emulsion of fine oil droplets from the respiratory passages and, as mentioned earlier, water projected into the air by the blow itself. Powerful muscles control fibrous nasal plugs that normally close the blowholes and fit into the internal nostrils when the muscles relax. The greater the hydrostatic pressure, the greater the closing off of the blowhole.

In order to breathe, whales have to voluntarily contract these muscles to extract the nasal plugs from the bony nares. In Mysticetes, the contraction of these muscles produces a V-shaped splashguard (or a blowhole crest) which prevents water from pouring into the blowholes during respiration.

The seal diaphragm is more like those of other mammals, but it does not prevent the lungs from collapsing completely during a dive due to hydrostatic pressure. The nostrils are kept closed by muscles. It is by a voluntary act that pinnipeds open them to breathe.

The breathing sequence is described as a series of respirations performed at short intervals followed by a normally longer period of apnea. A longer apnea period ends with a more explosive initial blow, higher and more powerful, followed by a breathing sequence involving more respirations, which is completed at even shorter intervals. As opposed to seals that exhale prior to a dive, cetaceans submerge with lungs that are filled with air. The amount of air they carry in their lungs is insufficient however to sustain the long apnea periods to which they are accustomed. Furthermore, the increasing water pressure compresses the lungs and forces the air into the part of the respiratory passages where no gas exchange is possible. As a result, other avenues of storing oxygen while at the surface and more economical ways of using it during a dive are needed.

Terrestrial mammals and cetaceans have in their blood, and in their muscles, a special molecule called hemoglobin that carries and stores oxygen.

Let's compare the way in which oxygen is distributed in the body of a human being with that of a cetacean.

	Total O_2 store in ml/kg	% in Lungs	Blood	Muscles
Phocids	60	07	65	28
Otariids	40	13	54	33
Cetaceans	35	22	30	48
Sea Turtles	20	52	42	06
Penguins	58	29	38	33
Humans	20	24	57	15

Significant differences are seen in the amount of oxygen stored in the lungs and in the muscles. But how can these differences be explained?

Lung volume: marine mammals have a proportionately smaller lung volume, a not surprising fact when we consider the constraints that hydrostatic pressure exerts on these animals.

Muscles: The muscles of cetaceans contain large amounts of muscle hemoglobin called myoglobin which allows for increased oxygen storage in muscles and gives these tissues their characteristic dark red color (nearly black in the sperm whale). This hemoprotein has an affinity for oxygen which is intermediate to that of the blood, which carries oxygen from the lungs, and to that of the electron transport chain or CYTOCHROME enzymatic system, which uses and stores oxygen before transferring it to the cytochromes. This additional amount of oxygen is, in itself, still insufficient to account for the long apnea periods of cetaceans. Other reasons also come into play: cetaceans have developed an increased tolerance to the presence of LACTIC ACID in their muscles, allowing them to perform longer under anaerobic conditions. In other words, they can produce energy without using oxygen. They will, however, accumulate a certain oxygen debt and, in addition, will at some point have to eliminate the poisonous lactic acid produced. The lactic acid is therefore oxidized and transformed into carbon dioxide (CO_2). This metabolic process takes place while the animal breathes at the surface. It explains why whales have to hyperventilate between submersions and why the breathing sequence is longer after prolonged dives: they have to eliminate their oxygen debt and replenish their oxygen store at the same time!

Retia mirabilia in the neck region of a Minke whale.

Artery and vein of the retia mirabilia in the neck region.

Video cameras attached to the back of dolphins to record the movements of their caudal fin have revealed that, at a certain depth (about 70 m or 230 ft), the pressure of the surrounding water becomes so great that the lungs are totally collapsed. Following this compression, the animal, denser, sinks like a rock to the "desired" depth, without any muscular effort. This reduction in energy expenditure significantly lowers oxygen consumption. Furthermore, when the animal has to get back to the surface, it simply has to swim back up to within 70 metres (230 ft) of the surface and let its lungs and thoracic cage expand once more. It can then float effortlessly back to the surface, as seen in a U.S. Navy movie on deep-sea diving experiments involving Tuffy, a trained dolphin.

Blood: Hematocrit (percentage of red corpuscules carrying O_2 / total blood volume) is greater in marine mammals, peculiarly in phocids. Red corpuscules are also bigger containing more hemoglobin. The total blood volume/body mass is 2 to 3 times greater than in non-diving mammals.

Whales have developed yet an another physiological response to apnea that we must consider. Although poorly known, it consists in the redistribution of blood during a dive. This response (which exists in the cetaceans and in the pinnipeds, such as seals and Walrus) must not, however, compromise the integrity of organs for which oxygen is essential. A mechanism that maintains blood flow to these organs must therefore exist. Consequently, during a prolonged apnea, body structures such as the brain, the spinal cord, and the myocardium, which cannot accumulate an oxygen debt, will receive a steady flow of blood, while the

Heart of a Harbour seal in situ. The lungs were slightly inflated. P: lung; C: heart; D: diaphragm

muscles and viscera, which already have their own provisions (oxygenated myoglobin), will receive almost no blood at all. Around the spinal cord and in the thoracic cage of cetaceans, there are complex networks of anastomosed arterial masses (retia mirabilia) that undoubtedly play an important role in this redistribution of the blood flow. They may also absorb the blood that comes from the thoracic cage, compressed as a result of the increased hydrostatic pressure during a deep dive. After having been distributed in the retia, the blood that has left the thoracic cage under great pressure can then reach the carotids with the appropriate pressure and be transported to the brain safely, preventing any damage to that organ.

During apnea, the heart rate diminishes considerably (bradycardy) so the blood pressure will be lowered in the parts where the blood still circulates.

Studies have shown that among phocid species capable of particularly long apnea, the body temperature can be lowered down to 35°C during long and deep dives. When the temperature is lowered in that way, metabolism rates are lowered too, so O_2 consumption diminishes and energy is saved.

The spleen is large, peculiarly among phocids. It acts as a reservoir of oxygenated red corpuscules which will be released as needed during the dive, furnishing a supply of O_2 to the organs and postponing the need to move into anaerobic energy production. There is a strong correlation between the size of the spleen and the capacity to do long apnea in phocids.

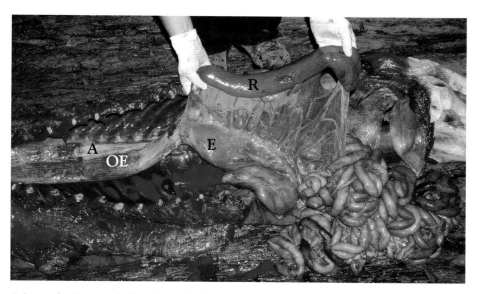

Spleen of a Grey seal (female and adult). A: aorta; OE: esophagus; E: stomach; R: spleen.

Spleen of a juvenile Harbour seal. E: stomach; R: spleen

All those adaptations allow marine mammals to make long apneas. It has been noticed, though, that marine mammals rarely dive longer than their Aerobic Dive Limit (ADL: the maximum diving time in which there is no augmentation of blood lactic acid concentration). The physiological adaptations mentioned above are efficient enough to allow them to meet their needs within their ADL. This dispenses with the need to have long periods on the surface to metabolize the lactic acid accumulated during very long dives. The Weddell seal which is able to stay more than one hour underwater rarely does so. In fact most of its dives are around 20 minutes which seems to be within its ADL, because at that rate it can dive almost continuously, without spending much time on the surface. After long dives, though (1h or more), it seems exhausted and spends several hours sleeping.

The ability to reduce body core temperature to lower metabolic rate during periods of apnea perhaps can explain why Elephant seals (*Mirounga sp.*) and Grey seals (*Halichoerus grypus*) spend a long time deep under water when they travel or forage. It is probably less costly to do that because they use less energy that way than in resting or moving on the surface. They can also shelter themselves that way from certain predators swimming nearer the surface (White sharks).

MARINE MAMMALS AND NITROGEN

Scuba divers are well aware of the dangers associated with nitrogen, specifically its toxicity and its supersaturation in the tissues.

TOXICITY

Air consists of about 80% nitrogen and 20% oxygen. When a diver breathes compressed air, the nitrogen in that air diffuses and dissolves in the blood and tissues in increasing amounts. Beyond a certain depth (which varies according to individuals, age, health, etc.), a diver may experience a kind of intoxication known as nitrogen narcosis or "rapture of the deep." This type of narcosis can manifest itself at a depth of 30 m (98 ft). The first signs consist in a feeling of increased disorientation accompanied by either euphoria or intense anguish. In such cases, a diver may possibly not feel the need to return to the surface or neglect to monitor air consumption or dive time, and eventually die.

TISSUE SUPERSATURATION

When liquids are subjected to increasing pressure, they can accumulate increasing amounts of dissolved gases. As a diver descends below the surface of the water, the hydrostatic pressure exerted on him increases considerably. In order for him to breathe without effort, he must breathe air whose pressure is equal to that of the water exerted on his thorax. His tissues, which consist of at least 70% liquid, begin to dissolve increasing amounts of nitrogen.

The air pressure in the lungs increases continuously with the rising hydrostatic pressure. The resulting compressed air remains in areas where gas exchange is possible, and nitrogen can then diffuse into the blood and, from there, to the tissues. When supersaturation attains

Cartilage rings around the bronchioles of a Blue whale.

Boat

Funnel to collect Tuffy's air.

Tuffy

Photo from a video of the U.S. Navy showing the technique used to collect the exhaled air of Tuffy, a trained dolphin.

We see the effect of the hydrostatic pressure on the dolphin's thoracic cage (below the pectoral fin). This picture was taken at 300 m (984 ft) by the U.S. Navy.

The thoracic cage is crushed by the hydrostatic pressure.

"critical levels," the diver must make decompression stages during the ascent to eliminate a certain amount of nitrogen before returning to the surface. This critical supersaturation level can be reached in six minutes by diving to a depth of 45 m (148 ft). A man performing successive dives to these depths by holding his breath (apnea) could also reach this critical level. As the diver returns to the surface, the hydrostatic pressure exerted on his body decreases and the nitrogen that has been absorbed by his tissues is released as bubbles. If no decompression stages for ascent are allowed, or if insufficient lung ventilation is performed between the dives, the nitrogen released from supersaturated tissues will form bubbles too large for their elimination via the airways. These bubbles can then end up in the general circulation and obstruct blood vessels anywhere in the body, including the heart and nervous system. Serious accidents can result, ranging from paralysis to even death. This condition is called "decompression sickness" (DCS), more commonly known as "the bends" or "caisson disease."

How do marine mammals, which perform repeated dives to depths greater than 1,000 metres (3,281 ft), avoid this critical supersaturation level and its disastrous consequences?

Remember that marine mammals submerge with very little air in their lungs (seals completely empty their lungs before diving). Furthermore, the particular positioning of their diaphragm allows the hydrostatic pressure exerted on the thorax and viscera to collapse the lungs. As a result, the pulmonary alveoli are emptied of air. The air is subsequently forced into the parts of the respiratory tract where no gas exchange with blood can take place. The smallest bronchioles of marine mammals are kept open by cartilage rings or strong muscles.

Some scientists, from studies of oil droplets in blowholes, believe that those droplets in the respiratory tracts of whales serve to fix, or trap part of the pulmonary nitrogen (nitrogen is very soluble in oil). The level of nitrogen in the tissues of whales remains, then, relatively constant during a dive, perhaps explaining their insensitivity to this gas. It therefore becomes unnecessary to imagine any miraculous methods, for example, the presence of bacteria in the blood capable of fixing nitrogen (referred by some as bacteria X, since they were never found). It is simply the amount of air that comes into contact with the blood that makes all the difference. Significant amounts of this air are in the scuba diver at ambient pressure, less air in the free diver (whose lungs are not adapted for a prolonged apnea) with ever-increasing pressure, but in whales, only minimal amounts are found and concentrated in the regions where no air-blood exchange can take place.

Experiments by Ridgway (1972) at the U.S. Navy laboratories in Point Mugu support these findings. He measured the amount of oxygen present in the air exhaled by Tuffy, a dolphin trained to blow underwater into a funnel connected to a container.

Following an apnea period of 3 minutes and 45 seconds, performed at a depth of 300 m (984 ft), Tuffy's exhaled air contained more oxygen than was collected from identical periods completed at 20 m (66 ft) and or just below the surface. Nearer to the surface, Tuffy could make use of the oxygen in its lungs, but could not do so at a depth of 300 m (984 ft) because the hydrostatic pressure had collapsed the lungs.

Recent studies being made on Sperm whale skeletons, as well as older studies found in museums, have shown bone damage which could be compatible with osteonecrosis related to decompression sickness. Cetaceans diving deep and long could not be immune either anatomically nor physiologically to the effects of too fast a return to the surface (like one provoked by the use of too loud sonars). (Michael J. Moore *et al.*, Dec. 2004)

DIVING ADAPTATIONS IN THE SPERM WHALE

The Sperm whale (*Physeter macrocephalus*) holds the record for deep dives. Sonars have detected individuals as far down as 2,250 m (7,382 ft). Food items recovered from one Sperm whale's stomach by a whaling ship included fresh specimens of bottom-dwelling sharks. The depth of water in the area was more than 3,000 m (9,842 ft).

Whalers had long noticed that when a Sperm whale submerged, it usually reappeared at about the same spot. Although this ritual facilitated their capture, it also shed light on this whale's hunting techniques and buoyancy mechanism. In fact, it seemed to show that instead of actively pursuing its prey, this animal waits in ambush to locate it with its biosonar (echolocation). If the Sperm whale did chase its prey, it would be hard, not to say impossible, for the animal to surface in the same location from where it submerged. To hang motionless, it must therefore have a way of controlling its buoyancy perfectly; otherwise it would be compelled to swim to remain in place, which brings us back to the "on-the-spot" surfacing problem. In 1978, Malcolm Clarke came out with an extremely interesting hypothesis on the role of the spermaceti organ during dives and in the buoyancy control of the Sperm whale. Some cetologists today challenge these hypotheses, but offer few arguments in return. Once again, we are reminded of the difficulties associated with studying these behemoths.

The head of the Sperm whale is huge, but its bones represent only 12% of its mass. The rest consists mostly of the spermaceti organ. Many years were needed for us to finally understand the exact anatomy of this organ. It was difficult to dissect the structure because it was enormous, and weighed perhaps up to 16 tons, the equivalent weight of three bull elephants! To learn how the spermaceti organ functioned, Clarke managed to convince a flensing man at a whaling station to slice the head of a Sperm whale. The man sectioned the huge head of the sperm whale into 20-cm (~8-in.) slices with the help of an enormous steam

Cross-section of a Sperm Whale's Head

Head of a Sperm whale. The case filled with liquid spermaceti is shown on the right; the junk is on the left.

saw. Clarke was then able to study the anatomy of the spermaceti organ in great detail.

The spermaceti organ is made up of two parts: the *case* and the *junk*. The case is a highly resistant fibrous container, a veritable case, or tank, containing up to 2.5 tons of oil known as spermaceti, recognized for its remarkable lubricating qualities. The case rests on blocks of spermaceti tissue, which are separated by thick fibrous septa that make up the junk.

As in all Odontocetes, Sperm whales have but one external nostril or blowhole. This apparent simplicity hides, in fact, an extremely complex arrangement of the nasal passages that starts at the base of the skull and extends to the single blowhole.

The somewhat tubular left nasal passage is essentially the sole breathing tube. It exits the skull by the left bony nare, which is much larger than the right one, runs along the left side of the spermaceti organ and ends up below the blowhole in the vestibular sac. The right nasal passage is however much different. It starts at the very small right bony nare and broadens to form the nasofrontal sac, which lies on the concave surface of the skull. From there, a flattened tube, more than one metre wide, passes below the case and ends in the vestibular sac. The case is therefore almost totally surrounded by these structures.

The advanced state of putrefaction allows the tendons of the maxillo-nasalis muscle to be seen.

Spermaceti, long believed to be the sperm of the Sperm whale, is a type of straw-colored oil, composed of waxes and triglycerides. Liquid at temperatures above 30°C, it tends to crystallize at lower temperatures. After having subjected different spermaceti samples to various temperatures and pressures, Malcolm Clarke found that the density of the spermaceti increased as temperature decreased. He thereupon suggested that the case served as a hydrostatic organ: all he had to do was find the way in which the Sperm whale could control the temperature of its spermaceti.

The architecture of the structures that surround the nasal passages help explain this temperature control. An extremely powerful muscle called the maxillo-nasalis muscle is found at the top of a Sperm whale's head.

As the maxillo-nasalis muscle contracts, the case is lifted and the blowhole opens. Water then penetrates the right nasal passage, which, we noted earlier, almost completely surrounds the spermaceti. As the cold water from the ocean cools the spermaceti, its density increases. Again, as we saw earlier, the blood circulating in the skin can accelerate this cooling. Consequently, the Sperm whale, having become denser than the water supporting it, sinks towards the bottom of the ocean (negative buoyancy).

If the whale needs to hang motionless at a certain depth, it simply has to warm up its spermaceti so that its density equals that of the surrounding water. The whale is then said to be neutrally buoyant. If the Sperm whale continues this warming process, the density of the spermaceti will decrease and positive buoyancy is reached. The animal then rises to the surface.

This is exactly what submarines do with their ballast tanks, or scuba divers with their buoyancy compensator. But instead of varying its density by blowing air into special compartments—which would necessitate, because of the hydrostatic pressure, huge quantities of air at those depths—the sperm whale simply has to relax its maxillo-nasalis muscle and adjust its blood flow around an almost incompressible reservoir. These effortless descents into the great depths (followed by equally effortless ascents) have the added advantage of contributing to the ways of reducing oxygen consumption and, in part, account for this whale's amazingly long apnea periods.

Skeleton of a Sperm whale including a model of the spermaceti organ.
British Museum

Clarke (1978) suggests that for a Sperm whale to maintain a depth of between 200 and 1,000 m (656 and 3,281 ft.), it would have to cool the 2.5 tons of spermaceti for only three minutes. In relation to this event, we note that a sperm whale takes about 15 minutes to descend a thousand metres and resurface, and spends usually more than one hour foraging at the desired depth.

Purves and Pilleri (1983) contest the role of the right nasal passage in the cooling of the spermaceti. They claim that seawater would enter this nasal passage and move the air inside it, making sound production necessary for echolocation impossible (see section on echolocation). In addition, the risk of introducing foreign bodies (pathogens) into the system would be too great.

They suggest that if the melon is indeed a hydrostatic organ, cooling of the spermaceti must be achieved by countercurrent heat exchangers described earlier or by a reduction of blood flow around the spermaceti organ. If so, the warming process would be carried by the vasodilation of the vessels that surround this reservoir and by increasing blood flow. Whatever the role of the spermaceti, it is clear that Sperm whales have, in their evolution, adapted remarkably well to life beneath the ocean. And their adaptations are all the more remarkable for having been developed from the organ functions of terrestrial animals.

Differences Between Phocids and Otariids

If seals are without doubt the champions in relation with depth and duration of their dives, fur seals and sea lions usually make shorter and shallower dives, even if they are able to go quite deep (more than 400 m for Steller sea lion (*Eumetopias jubatus*) or California sea lion (*Zalophus californianus*). Maximum measured time underwater for those species is 15 and 16 minutes. Their lungs do not collapse completely during a dive; their O_2 carrying capacity is intermediate between large phocids and terrestrial mammals, and their spleen is not much larger, as a percentage of body weight, than in terrestrial mammals. Compared to seals their blubber is thinner and their fur thicker. Their shape is a little more hydrodynamic and maybe, they put more emphasis on speed rather than long dives to capture their prey, even if energy costs are higher.

CIRCULATORY SYSTEM

The circulatory system of marine mammals is quite similar to the standard mammal system. Yet, their completely aquatic way of life and the long apnea periods for feeding create constraints that have resulted in special adaptations.

Section of a cetacean heart. There are no differences in the thicknesses of the two ventricles.

The hearts of cetaceans, even if they look a lot like other mammals' hearts, show less difference in the thickness of the walls of the 2 ventricles. In land mammals, the walls of the left ventricle, which has to push the blood in the whole body, often against gravity, is very muscular, a lot thicker than the right ventricle which has only to circulate the blood in the lungs. Living only in water where they are supported by the medium, does not impose the same stress on the cetaceans that land mammals have to endure. One has only to think at the work a giraffe heart has to do to push up the blood in the head!

Heart of Minke whale.
The Îlets-Jérémie 2000.

Heart of Sperm whale. Its size is like a 100 litre barrel.

Position of the heart at the White-sided dolphin. The heart is placed completely ahead in the rib cage, probably to facilitate the collapsing of the lungs at the time of the major apnea.

Heart of the Blue whale.

Blue whale, sigmoid valve.

The Blue whale's aorta.

59

The heart is placed completely forward in the rib cage, no doubt to allow for the easier collapse of the lungs during deep diving.

Pinniped hearts are, on the contrary, very much like those of land mammals because they have to function, at least part of the time, under the same conditions. Their position is not very different of what we find in land mammals.

The rest of the circulatory system is not very different. In addition, we find very extensible blood vessels that act as blood reservoirs (rete mirabile) and, at least in pinnipeds, a sphincter on the inferior vena cava which limits the circulation of the blood during apnea to the heart and the central nervous system.

Heart of the Harbour seal.

Transverse section of the Harbour seal's heart. Notice the large difference between the right (V.D.) and left (V.G.) ventricles.

PERCEIVING THE ENVIRONMENT

It is the quantity and the quality of the information gathered by an animal in its environment (i.e., its sensory capabilities) that determine its adaptability and competitiveness. It is difficult for humans, who act in response to an almost exclusively visual representation of their environment, to imagine the world of a dog, an animal that relies mostly on its olfactory or auditory capabilities, or that of a whale, whose world is almost entirely one of sound.

The sense organs are the windows to the outside world. They supply the central nervous system with the necessary information to ensure the survival of an organism in an often-hostile environment in which it must preserve its integrity as well as find food. Although their ancestors were terrestrial, cetaceans have adapted to living in an aquatic environment. Consequently, changes must have occurred that permitted whales to successfully compete in their habitat. As predators, whales had to improve specific senses to be able to function efficiently in their new environments; those of little value have lost much of their acuity (or have completely disappeared!).

TOUCH

CETACEANS

The study of the nervous system of cetaceans reveals that their sense of touch, as is the case for most living creatures, gives them access to an important source of information about the outside world. In fact, areas of the whale brain responsible for collecting and analyzing tactile information are well developed (the *lobulus simplex* for the head and the *paraflocculus* for the trunk and caudal fin). The trigeminal nerve (one of the largest cranial nerves) and other nerves involved in the transmission of tactile information from the trunk and caudal fin are also highly developed.

The skin of cetaceans is very sensitive, as is known by people fortunate enough to have touched them. In fact, a simple touch brings on an immediate reaction every time. We have seen how important it is for the skin to react instantaneously to point pressure variations in order to maintain a laminar flow over the body. (Whether this phenomenon does indeed exist remains to be validated, however.) Be that as it may, many cetaceans in captivity and sometimes others in their natural habitat seem to enjoy being rubbed or scratched, an indication that the skin is a site of pleasurable sensations.

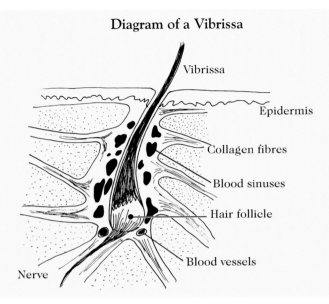

To increase tactile sensitivity, there exist, mostly in Mysticetes, vibrissa-like hairs at the tip of the chin, over the lips, and around the blowholes.

A vibrissa is a stiff hair whose base is embedded in a hair follicle. Any hair movement will stimulate the nerve endings at its base and amplify sensory (tactile) perception. The vibrissae of cetaceans lack ARRECTOR

Vibrissa at the tip of a Minke whale's lower jaw (île Verte, October 1996).

PILI MUSCLES, but their base is surrounded by capillaries that possibly set these hairs upright when they are filled with blood. Far from being strictly for decorative purposes, vibrissae are most certainly useful to feeding whales.

The Humpback whale is one of the "hairiest" cetaceans. Its hairs, more than a hundred, are found in the tubercles distributed in rows throughout the head and at the tip of the chin.

Vibrissae seen here in the head tubercles (or sensory nodules) of a Humpback whale.

Yves Poirier

Why have baleen whales kept hair, which is almost totally lacking in toothed whales? Vibrissae may allow baleen whales to assess the density of the PLANKTONIC animalcules and thus save valuable energy by not gulping and sieving low prey-density waters. In Odontocetes, only the hair follicles remain and have been modified to form highly innervated furrows. According to English cetologists Palmer and Weddell, as mentioned by Tomiline (1974), these furrows may serve to assess their speed. Among Odontocetes, vibrissae are found only on the snout of River dolphins (*Platanistidae*) and are used to locate food in the muddy or murky waters in which they live, a habitat where the sense of touch is, to all evidence, essential (Tomiline, 1974).

Ganges River dolphin (*Platanista gangetica*)

63

PINNIPEDS

Pinnipeds are equipped with very long vibrissae. Those vibrissae are placed in 3 patches on the head.

Mystacial and supraorbital vibrissae (phocids and otariids) and rhinal vibrissae (phocids only) They are oval and smooth (otariids, Walrus, Monk seals [*Monachus spp.*] and Bearded seal [*Erignathus barbatus*]) or very flat and with wavy surface (all other phocids) They are very richly innervated and sensitive in all mammal species. Pinnipeds' vibrissae are surrounded by 10 times the number of nerve fibers and sensory cells (Merkel cells) found in other well endowed mammals like the rat or the cat.

Vibrissae of the Harp seal.

What are they used for? They can be used to identify size, shape, and texture of objects, a good means to identify prey.

They can be used, too, as hydrodynamic receptors, to follow fish in order to catch them. When a fish swims, it creates pressure waves in the water with its tail movements. Those pressure waves remain for some time in the medium. It was shown that a seal can identify and follow the "pressure track" left by a fish in the water, even if the fish is 40 m distant That would explain how seals remain efficient predators in murky water or at a depth where there is no light. Seals' vibrissae are as sensitive to pressure as lateral line organs in fish. In order to maintain sensitivity, the skin areas in which vibrissae grow has to be maintained at core body temperature, even if it causes a great energy expenditure, so important are they in the life of the seal.

The skin also has many nerve endings associated with particular sensory functions such as MECHANO- and BARORECEPTORS (bulbs of Krause and Pacinian corpuscles).

Little is known about the perception of temperature variations in marine mammals. The fact that some species confine themselves to waters that have relatively narrow temperature ranges (5 to 20°C, for example, in Right whales), however, suggests that they are sensitive to this type of fluctuation.

Is pain, associated with the sense of touch in humans, part of the sensory perceptions of marine mammals? Unquestionably so, as it is for all animals. Any nerve impulse exceeding a certain intensity level will be perceived as painful. Organisms are thus able to react (i.e., flight or fight) to conditions that could endanger their well-being or homeostasis. Pain is therefore essential; it serves as an alarm signal. But to think that pain has the same significance for whales that it has for humans, who are endowed with *imagination*, reflects an ANTHROPOMORPHIC attitude. We have only to consider the apparent "stoicism" of a wounded animal to see that it can easily live with an injury that would be debilitating for us. We cease any potentially harmful activity from the *anticipated* fear of the pain that would follow. An animal will avoid pain if possible and even certain actions that are clearly associated with it. But an animal will also tolerate pain perhaps without ever becoming aware that it is unable to do otherwise. Films of oxpeckers pecking in the opened wounds of large African ungulates, such as the giraffe, show exactly that. Oxpeckers feed on insect larvae that hatch from the eggs laid in the wounds. By doing so, these birds prevent healing and keep the wounds opened. Their acerated beak scouring through the wounds must induce sensations that are surely more painful than pleasurable. All the same, the giraffe does not react and seems nobly indifferent to the whole situation.

The sense of touch is therefore well developed in cetaceans. It plays an important role in friendly social contacts (mother-calf, adult-adult) and during courtship periods. It also serves to establish hierarchies among the individuals of a group, as evidenced by the visible scars (tooth marks for example) on the sides of male Odontocetes. Touch plays a vital role in the overall image whales make of their world: similar to a blind man who discovers his surroundings through his finger tips, some cetaceans use the richly-innervated tip of their penis to "feel" their environment. It may even be possible that their highly sensitive skin and the presence of vibrissae are used, at least in Mysticetes, in sound perception (at least low frequency sounds). These sounds may be perceived as pressure waves, or mechanical stimuli, mostly by the anterior portion of the head.

OLFACTION

The areas of the brain specialized for perceiving and analyzing olfactory information are greatly reduced or, more often, non-existent.

Mysticetes have kept their OLFACTORY EPITHELIUM which links up with the olfactory brain centers via cellular processes that pass through the lamina cribrosa (cribriform plate) of the ETHMOID.

This longitudinal section shows that the lamina cribrosa of the ethmoid of this young Minke whale is perforated by many cribriform foramina, indicating that a connection exists between the nasal cavities and the braincase, through which the nerve endings of the olfactory nerve may pass.

Cranial cavity

Connecting bone of the nasal cavities (ethmoturbinates—scrolls of bone that increase the surface area of the olfactory epithelium).

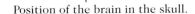

Position of the brain in the skull.

Longitudinal section of the cranial cavity of a Fin whale. The pointer rests on the lamina cribrosa in the area of the olfactory bulb, which is quite large in this species.

Position of the brain in the skull.

When considering the reasons Mysticetes have kept their olfactory mucosa, it is worth noting that whalers as well as biologists have noticed that plankton swarms and fish shoals give off a discernible odor when near the surface. Perhaps this odor can be detected by the plankton-consuming baleen whales during their breathing sequences. Its presence may allow for a more rapid detection of krill swarms. Might this explain why Mysticetes spy hop, that is, rise relatively straight up out the water and maintain their head above the surface?

Inside the skull of a Beluga whale. We see evidence of the ethmoid's lamina cribrosa. It is not perforated. The posterior half of the skull has been sectioned.

Odontocetes lack both olfactory mucosa and olfactory nerve. It is probably because their prey (fish or squid) do not give off odors at the surface that the pressures of natural selection, through Odontocete evolution, did not favor the maintenance of a functional olfactory organ. Odontocetes may give off underwater odors, however, which may explain why they, along with the Mysticetes, have kept the vomeronasal organ (or organ of Jacobson).

Vomeronasal organ of a Fin whale, Pointe-au-Père, October 1999.

The vomeronasal organ gives an animal the ability to "smell" its food once it is in its mouth. It is also used by animals such as the *Cervidae* (deer family) and the *Felidae* (cat family) in the detection of the PHEROMONES secreted by rutting females. In cetaceans, however, its role remains purely speculative.

Pinnipeds have kept a good sense of smell, even if their olfactory bulbs are relatively small. The sense of smell may be used by female to recognize their pups.

TASTE

The great number of nerve endings (taste buds) in the tongue of cetaceans as well as the size of the nerves and areas in the brain responsible for gathering and analyzing gustatory information indicate that the sense of taste is as developed in whales as in other mammals. Some experiments with captive dolphins showed that their sensitivity to "acid" (citric acid) or "salty" is tenfold what it is for humans. Cetaceans probably use taste to evaluate the salinity of water and to detect the presence of other whales (feces, urine), or, maybe, to recognize by taste certain areas of the ocean, near big rivers, where large salinity gradients exist, which would enable them to localize themselves on their "internal map" and recognize more productive areas. By contrast their sensitivity to "sweet" is 10 time less than in humans. Much remains to be learned about their use of gustatory information, which may make for their lack of sense of smell.

Tongue of a Minke whale.

VISION

Contrary to land-dwellers, cetaceans live in a world where little light is available. In fact, water reflects and absorbs light. The smaller the angle at which light strikes the surface of the water, the more light will be reflected. Consequently, light penetration will be maximal when the sun is at its zenith, and will decrease as it approaches the horizon.

Light penetration is also affected by the amount of suspended particles in the water (turbidity). The visible portion of solar radiation is composed of various wavelengths ranging from red to violet. Even in the clearest of waters, red is completely absorbed at a depth of 10 m (33 ft), orange and yellow disappear between 10 and 60 m (33 and 180 ft), green and blue between 60 and 140 m (180 and 450 ft).

Close to 90% of light's energy is absorbed at this depth, and at 200 to 400 m (656 to 1,312 ft), depending on the turbidity level, there is no more light penetration. Horizontally, in the best of conditions, one must not expect to see more than a distance of 60 m (197 ft). With this in mind, it is safe to say that vision in cetaceans may not be their most utilized sense, especially in those that feed on minute prey or swim at great depths.

The cetacean eye is not fundamentally different from that of a terrestrial mammal. It is relatively small: the eye of an eighteen-meter fin whale is about the size of an orange. If it were proportional to the human eye, it would be the size of a basketball.

Absorption of Light Energy in Water

Depth at which there remains only 1% of the light that has entered a clear ocean surface, according to its wavelengths.

The eye shows particular adaptations for seeing in a dimly-lighted underwater world. The interior chamber is oval to allow clear vision in an aquatic environment. The sclera is extremely thick, especially in Mysticetes, in all likelihood, to maintain this oval shape underwater to resist the hydrostatic pressure that would tend to make it spherical.

The iris can open wide to allow the weak underwater light to stimulate the photoreceptive cells of the retina. It can also close itself off, leaving only a narrow slit. This is done to reduce light penetration in the eye when it is intense (near or above the surface) and to increase depth perception. As in most aquatic animals, the lens is relatively spherical. Since the lens is very weakly elastic, whales obviously must have other means to adjust their vision, means that are still unknown to us.

The retina consists almost exclusively of RODS, which are photoreceptive cells adapted for scotopic vision (vision in low levels of light). Most of them are indirectly connected to the optic nerve, which also increases, through summation, the sensitivity of the retina in a murky world. These cells are adapted to monochromatic vision. Photoreceptive cells, or CONES, adapted for photopic vision (vision in bright light), are scarce. The fact that they are also associated with color vision suggests that whales are probably completely color-blind just as many other mammals are, or that their eye is only able to perceive certain colors. A Cetacean's retina is most sensitive to wavelengths corresponding to the blue-green part of the solar spectrum. It is reasonable if we consider

Eye of a fin whale.

The almost-spherical lens of a dolphin.

Eye of a Minke whale (île Verte, October 1996) The tapetum lucidum shines through the slit-like pupil (this type of pupil is also seen in the ungulates).

that the red orange part of the spectrum is very rapidly absorbed by the water. A TAPETUM LUCIDUM ("shining carpet"), a reflective layer composed of numerous guanine crystals covering the back of the eye, reflects unabsorbed light back to the retina, which enhances its sensitivity to light.

Many nocturnal or crepuscular animals have this reflective layer. It is what causes the eyes of a cat or deer to shine when caught in the headlights of a car at night.

There are no lacrimal (tear) glands and associated ducts, but there are glands that secrete a type of oil that lubricates the eyeball. Eye movement is possible and the eyelids can open and close. The visual range of cetaceans extends to the sides. Binocular vision is limited in most species, except in the smaller ones. Sperm whales and the large Mysticetes are almost capable only of monocular vision.

Cetaceans have relatively good aerial vision. We often see them spy hop, that is, raise their head relatively straight up and out of the water. Porpoises and dolphins are known for their very accurate high jumps to get food from the hand of their trainers.

Killer whale jumping in an aquarium.

François Légaré

The necessary aerial or underwater visual adjustments are probably affected through the combined action of the iris (which dilates the pupil under water and contracts it above the surface) and of changes in the curvature of the cornea (Tomiline, 1974). It has been shown, at least in Harbour porpoise (Phocoena phocoena) that the curvatures of the anterior and posterior corneal surface are different. That should compensate for the myopy in air of an eye adjusted to see clearly in water. Although vision is relatively unimportant for capturing food, it certainly plays a role in gregarious cetaceans and in those that show contrasting color patterns on certain areas of the body. Changes in the position of a pod member are thus easily detected by others and group movements rapidly synchronized.

Tapetum lucidum, optic nerve and pupil of the eye of a Harbour seal.

The pinniped's eye, like the cetacean's, must be able to see underwater and on land. It is relatively very big, with a tapetum lucidum, a pupil which can open wide and close almost completely vertically, as in cats or foxes. The cornea cannot be used to accommodate the vision, because with the encased fluids it has the same refractive index as sea water. It will be the eye lens, almost spherical, that will do it. When in air, the eye becomes myopic. The cornea and the fluid behind will be able to adjust the vision because the curvature of the anterior and posterior of the corneal surface are different, as in cetaceans. This, and the fact that the pupil can close almost completely, will allow the eye to function in air. As we see accommodation is probably made by the same means as in cetaceans.

The pinniped's retina has its maximum sensitivity in the same range of wavelength as the cetacean's. Coastal species, which do not dive deep, are more sensitive to green; deeper foraging species are more sensitive to blue. One can see the same kind of adaptation in fishes according to the depth at which they are usually found.

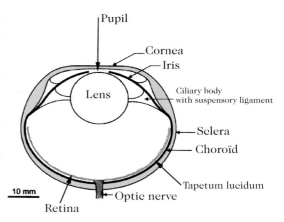

HEARING

The ear is the auditory organ. It is responsible for converting vibrational energy of the outside world, collected by specialized structures, to nerve impulses which are analyzed by the auditory centers in the temporal lobes of the brain.

The remarkable development of the acoustic nerve and of the regions of the temporal lobe clearly indicates that the sense of hearing is highly important in cetaceans.

How does the ear of a terrestrial mammal, such as a human being, function?

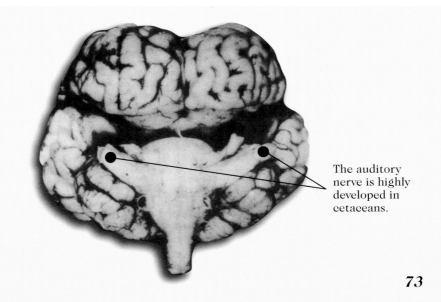

The auditory nerve is highly developed in cetaceans.

Airborn sounds are transmitted as vibrations. These vibrations are funneled by the auricle (pinna) down the auditory canal to the eardrum (tympanum). The latter is a membrane that stretches across the auditory canal and that begins to vibrate harmoniously with airborne vibrations. As the tympanic membrane vibrates, a chain of three auditory ossicles (malleus or hammer, incus or anvil, and stapes or stirrup) in the middle ear—located in the squamosal (or temporal bone)—amplifies and transmits these vibrations to the liquids in the LABYRINTH of the inner ear—found in the part of squamosal called the petrosal bone. In another part of the labyrinth called the *cochlea* are sensory cells that rest on a membrane. The vibrations of the liquids of the inner ear cause this membrane to vibrate and stimulate these cells. Nerve impulses are then sent to the auditory centers of the temporal lobe where they are interpreted as sound.

Sound Transmission and Reflection in Terrestrial and Aquatic Mammals

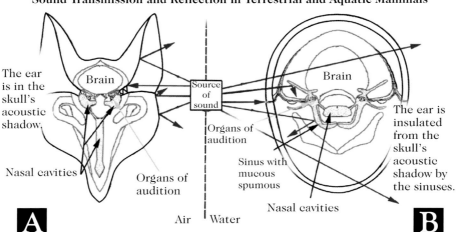

Diagram of sound transmission and reflection in terrestrial (A) and aquatic (B) mammals.

We can pinpoint the origin of a sound because airborne vibrations are asynchronous; that is, they do not reach our two ears at the same time. Although this difference seems insignificant (0.4 milliseconds when the source of sound is at 45° angle from our head), it is enough for our auditory center to locate the sound's point of origin.

Air has an ACOUSTIC IMPEDANCE that differs from that of the head tissues. As a result, 90% of the sounds hitting our head are reflected. The auditory canal is then the only avenue by which sound can be converted into nerve impulses, and we always have one ear in the "acoustic shadow" of the head with respect to the other and to the source of the sound.

Things are quite different in water. Upon immersing our head, we are amazed by the general increase in sound intensity. We also lose all notion of sound depth. We are no longer able to locate sounds because the acoustic impedance of the head tissues and that of the water are almost the same. Sound waves therefore travel through the head and through water at the same speed. Consequently, the vibrating bones of the head "harmonize" the movements of the liquids in the ears. The acoustic shadow of the skull on either one of the ears no longer exists. Nerve impulses leave the ears at the same time. The auditory centers of the brain, which are simultaneously informed by the two ears, are no longer able to maintain depth of sound.

The auditory system of whales has to be different. It must be able to locate underwater sounds, those possibly associated with orientation and the detection of possible dangers and food (see the chapter on echolocation). In order for a whale to use underwater sounds effectively, that is, to conserve its binaural hearing, it has to acoustically insulate its two ears from one another as well as from its skull.

The work of Fraser and Purves (1954-1960) has shed considerable light on the anatomy and physiology of the cetacean ear. In human beings, the auditory organs used for sound perception (phonoreception) and STATORECEPTORS (or equilibrium receptors) are embedded in the temporal apophyses of the skull, namely, the *petrosal bone* (Gr. *petros* meaning stone, a very hard bone). The squamosal (temporal bone) is fused with the rest of the skull. In cetaceans, however, it is no longer the temporal apophyse that holds the auditory organs, but a distinct bone called the *petrotympanic bone*. This latter consists of the *petrosal bone* and the *tympanic bulla* (or the auditory bulla) and is not fused with the rest of the skull, but is, instead, loosely connected to the squamosal by the mastoid apophyse. In Odontocetes, a small but thick and ligamentous cushion connects this apophyse to the squamosal. In Mysticetes, the relatively much longer mastoid apophyse is loosely inserted between the squamosal and the occipital bone.

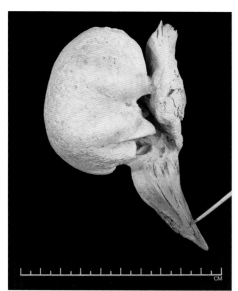

Petrotympanic bone of a Mysticete (Minke whale); the pointer shows the mastoid apophyse. The tympanic bulla is on the left.

The petrotympanic bone, particularly in Odontocetes, is completely insulated from the rest of the skull by complex sinuses (the periotic sinuses, among others) that are filled with dense foam of gas bubbles in an emulsion of oil and mucus. This foam is extraordinarily resistant to pressure. Experiments have shown that its volume remains practically unchanged even when subjected to pressures of 100 atmospheres (the equivalent of a depth of 1,000 m or 3,281 ft!).

If it does experience a decrease in volume due to the increasing pressure, a large number of blood vessels in the sinuses dilate to maintain structural integrity and prevent fractures during deep dives.

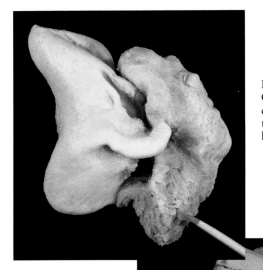

Petrotympanic bone of an Odontocete (White-sided dolphin); the pointer shows the mastoid process. To the left is the tympanic bulla.

Beluga's petrotympanic bone in the periotic sinus.

The millions of gas bubbles contained in this foam give it an acoustic impedance similar to that of air. The sound vibrations reaching the head and transmitted by the cranial bones are either reflected or absorbed, leaving the whale's inner-ear organs undisturbed. These organs are thus acoustically very well insulated.

Spumous mucus in the periotic sinus of a Minke whale. On the right is the tympanic bulla.

Beaked whales (Odontocetes of the *Ziphiidae* family) are unique in that the mastoid apophyse of the petrotympanic bone is solidly inserted between the squamosal and the occipital bone. We may ask ourselves if the acoustic insulation of their ear structures has attained the level of specialization seen in other Odontocetes, in which the use of sonar is crucial to perceiving their world. Oddly enough, a poorly insulated auditory system would surely be a handicap for these cetaceans that generally feed on single prey items (squid, herring, etc.). Since beaked whales do not seem to suffer such a handicap, we must assume that this structure is more poorly insulated in appearance only. A similar situation is also seen in the *Physeteridae* family, in which the petrotympanic bone is fused with the rest of the skull, but to a lesser degree.

The only avenue by which soundwaves can stimulate the ORGAN OF CORTI is through the normal pathway, that is, the auditory canal, which includes the eardrum, the ossicles, and finally, the liquids of the inner ear. In addition to having an extremely small external auditory canal, the external auditory meatus, or passage, is often blocked by fibres or an earwax plug.

Petrotympanic bone of a Northern Bottlenose whale *in situ*. The mastoid apophyse is highlighted in red.

Experiments on neighboring tissues have revealed that the auditory canal is as effective as that of terrestrial mammals. Each ear functions independently of the other (binaural hearing). In addition, although sound travels faster in water than in air, the head of a cetacean is large enough to create the necessary asynchronous sound reception (acoustic shadow effect) allowing for binaural and ultimately directional hearing. Another hypothesis, however, suggests that sounds reach the organs of the middle ear via the tympanic bulla, set in motion by the vibrations in the water and conducted by the fat-filled cavity of the lower jaw of Odontocetes.

Auditory canal of a Minke whale.

Finger-glove eardrum of a Minke whale. The malleus or hammer (encircled portion) is attached to it. To the right is the unblocked auditory canal. There is no wax plug because the animal was too young (probably in its first year).

Auditory ossicles of various cetaceans.

In cetaceans, particularly in Mysticetes, the eardrum extends out of the tympanic bulla into the auditory canal forming a structure called the "finger glove."

A rather curious accumulation of KERATINIZED cells (skin particles) and cerumen (earwax) in the auditory canal of cetaceans, especially in Mysticetes, is seen to form an earplug. Since earwax layers accumulate over the years, scientists have attempted longitudinal studies of their growth to determine the age of Mysticetes. In the fin whale, data obtained by the Japanese (1965-1976) in Antarctica's Zone VI and cited by S. Mizroch, show that the layers are laid down on a biannual basis (growth layer groups).

General view of the two ears. From the right to left, the tympanic bulla (1), the occipital condyles (2), the sinus (3) and turned over tympanic bulla (4) are seen.

Close-up of the ear of a Minke whale. The tympanic bulla (1) has been removed to expose, from left to right, (2) finger glove, (3) malleus, (4) incus, and (5) stapes in the oval window.

It is extraordinary to see the evolutionary changes that took place to recover, transform, and fashion an aerially well-adapted structure into a highly functional aquatic acoustic marvel.

Ursula Tscherter

COMMUNICATION AND ECHOLOCATION (AUDIODETECTION)

Cetaceans emit an incredible variety of sounds. The frequencies of these sounds include those perceived by the human ear, that is, from 20 Hz to 20 kHz, but also include high-frequency sounds (supersonic), up to 250 kHz, as well as very low-frequency sounds (infrasonic) of less than 20 Hz. Whale sounds range from moans, trills, grunts, squeaks, creaks and whistles to series of supersonic clicks.

Why do these animals make so many different sounds?

The sounds emitted by Odontocetes (including dolphins and porpoises) are very different from those of Mysticetes (rorquals, Right whales, and the Grey whale). Odontocetes are, with the exception of the Sperm whale, generally much smaller than most baleen whales. They are usually gregarious and travel in large pods, or groups. The sounds they produce in the human audible range, mostly whistles, most certainly play a role in social communication. They are produced during rapid pod movements, upon arriving in a familiar territory, and when individuals that had wandered off from the group return. In short, they remind us of the chattering of moving monkeys, or of migrating geese. These sounds probably serve to maintain cohesion within the group.

Certain sounds are used to call for help. Whalers sometimes deliberately wounded a calf to induce the grouping of individuals, mostly females, attracted by its distress calls. The positive reaction to these calls probably plays a role in the mass strandings of gregarious species (see the chapter on strandings).

Sounds made by predators, such as the Killer whale, cause panic among seals and smaller cetaceans, and were imitated by humans, with limited success, to scare those same animals away from fishing nets.

Some gregarious species, such as Sperm whales, Narwhals, Killer whales, and other dolphins, have a vocal repertoire specific to each group which serves as a sound signature for individual identification. It is very likely that research in this field will reveal that this is indeed the case for all Odontocetes, seeing that they all have a complex social structure.

Series of clicks produced by two Sperm whales.

In addition to the sounds audible to the human ear, Odontocetes produce successive bursts of ultrasonic clicks. Each click is of short duration (ranging from a fraction of a millisecond to 2.5 milliseconds) and of frequencies that vary between 25 and 250 kHz.

These series of clicks are produced to locate prey or obstacles. As the sounds bounce off objects, the whale's specialized brain centers analyze the returning echoes. By comparing these echoes with the sound images in its memory, a whale can identify them. An appropriate response is thereby elicited depending on the nature of that image (i.e., attack or avoidance).

The method used by animals, such as bats, to locate obstacles or prey by means of directionally emitted sound echoes is called echolocation (biosonar). These high-frequency sounds—beyond the range of human hearing—bounce off objects and return to the ears of the echolocating animal. The appropriate centers of the brain subsequently analyze these sounds and an acoustic image (mental map) of the animal's surroundings is made. In Odontocetes, which must capture individual prey items, the image is understandably comparable to, if not superior to, those obtained in the best diagnostic ultrasound examinations, given that these whales must be able to accurately locate their prey in order to catch them on their first attempt.

Whales' ability to discern different objects is astonishing. The Bottlenose dolphin (*Tursiops truncatus*), the species popularized by the television series "Flipper," can differentiate between a 6.35 cm (25.5 in.) diameter steel ball and one having a diameter of 5.08 cm (2 in.).

Echogram (sonogram) of a human fetus (Justin, my grandson). In spite of its quality, it fails to compete with the sharp acoustic image of the Odontocete biosonar.

It can also locate a 7.62 cm (3 in.) diameter metallic sphere filled with water at a distance of 70 m (230 ft)!

The orca (Killer whale) can locate an air-filled plastic ring having a diameter of 10 mm (0.4 in.) at a distance of 3 m (9.8 ft). The Pacific White-sided dolphin (*Lagenorhynchus obliquidens*) obtains similar results, but with a 2 mm (0.08 in.) diameter disk! A dolphin can most certainly detect a school of fish at 100 m (328 ft), and even a 13 cm (5 in.) fish at a distance of 9 m (29.5 ft) and swimming away at a right angle.

When dolphins explore their surroundings, the rate at which low-frequency clicks are emitted is relatively slow. This is done in order to get a longer sonar range and to allow echoes from distant objects to come back. But, even though low frequencies travel farther, the acoustic image is of a lesser quality. When the target is detected, the rate and frequency at which clicks are emitted increase. The higher frequency allows for sharper acoustic images but reduces the range.

When they echolocate, cetaceans produce very short duration clicks whose frequency and rate are continuously adjusted, not only in function of the range but also to compensate for ambulant noise, so as to maximize the return of echoes containing pertinent information about the target.

As for Mysticetes, their lower-frequency sounds would suggest a cruder ability to echolocate—it is still uncertain whether or not these whales have sonar. Its function is at any event not significant for these whales, whose prey live in large, if not huge, groups.

Where do whale sounds come from? Once again, as is the case for many phenomena regarding these creatures, our knowledge is limited. Opinions about this subject are therefore varied if not contradictory.

The most current theory on how whales produce and perceive sounds in general, as well as those used in echolocation, comes from Norris, and is expanded upon by Ted Cranford. According to this theory, sounds emitted by cetaceans, particularly Odontocetes, are produced in the perinasal sacs, the widened area of the respiratory passages between the skull and the blowhole. Through the combined work of a series of muscles and of membranous folds acting as valves that surround these sacs, sound is made. Ultrasounds are produced just below the blowhole, in the diverticula of the nasal meatus by what Ted Cranford calls the monkey lips/dorsal bursae complex, or MLDB. The sounds used by the Odontocetean biosonar are created in the nasal meatuses by the movement of a structure (the anterior dorsal bursa), which hits a surface on the opposite side of the meatus (the posterior dorsal bursa). The latter, whose acoustic impedance is similar to that of water, then transmits sounds through the low density oil of the melon.

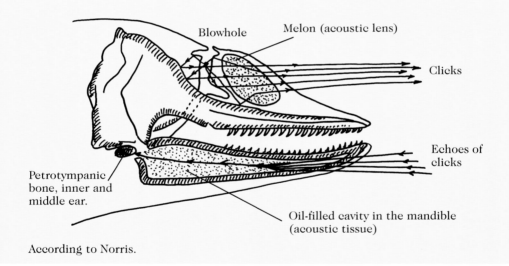

The Emission and Reception of the Sonar Clicks of a Dolphin

According to Norris.

This system is pneumatic, that is, it is set in motion by the air that comes from the larynx. Echoes from these ultrasounds, after reflecting off obstacles, return toward the animal. Upon reaching the animal, they are conducted directly to the auditory organs by the oil-filled cavity in the thin but dense bone (the acoustic window) of the lower jaw (mandible) which is attached to the tympanic bulla. The chemical nature of the oil contained in the melon is similar to that found in the lower jaw. These structures are often referred to as acoustic tissues, and the melon, in particular, as an acoustic lens.

This hypothesis, in spite of its popularity, is still highly controversial. Its conclusions are based on anatomical considerations particular to Odontocetes and on structural homologies (MLDB) in that group. With a critical eye to this theory, we note that the tissues of the melon, in particular, seem to lack the necessary refractive power to act as an effective acoustic lens, especially if sounds are emitted in the nostrils: the source of sound production seems too close to the melon for it to work this way. Finally, many cetaceans lack these structures but are perfectly able to echolocate. In addition, the convex-shaped skull of many Odontocetes makes it a bad acoustic reflector.

Nevertheless, recent studies seem to confirm that the nasofrontal sacs, particularly in the area around the nasal plugs, are the site of sound production in Odontocetes. By inserting a camera in the blowhole of an Odontocete in the process of emitting sounds, Ted Cranford (1988) got remarkable images that demonstrated a synchrony between the vibrating nasal plug and the emitted sounds. X-ray films also show movement in the area of the nasal plug.

Vocal cords in the larynx of a Bottlenose dolphin (*Tursiops truncatus*).

A second hypothesis states that all sounds, or at least some, are produced in the larynx as in terrestrial mammals. The work of Joy Reidenberg has shown that the larynx of Odontocetes bears structures analogous to the vocal cords of land mammals (Reidenberg *et al.*, 1987, 1988, and 1994).

The experiments by these cetologists suggest that whales' larynx structures can vibrate in the ranges necessary to account for the vast array of sounds produced.

Some cetaceans, such as the Beluga and the Platanista or Susu River dolphin, can also emit lower sounds (Zbinden *et al.*, 1979).

The structures that produce sounds are acoustically well insulated from the rest of the head, and especially from the auditory organs, by sinuses that are filled with mucus laden with microscopic gas bubbles. The whale, unable therefore to hear the sounds it produces during echolocation, adjusts its voice in response to echoes received, rather than from the emitted vibrations. If this adjustment to the frequency and range of the echo is to occur, it is important that no sound connection exist between the "transmitter" and "receiver" (Purves and Pilleri).

In this hypothesis, part of the echoes of emitted sounds would simply be received by the auditory meatuses, or canals, in the conventional mammalian fashion. Research has shown that if the "acoustic window" of an Odontocete is occluded by a patch of neoprene rubber, it will lose only part of its hearing ability. It is possible that high frequency sounds (biosonar) are received by the jaw and that lower frequency sounds (social vocalisations) are received via the auditory canal. It is the high degree to which the two ears are acoustically insulated from each other that allows directional hearing. Only the sounds that reach the auditory meatus or the acoustic window in its vicinity are transmitted to the auditory organs on either side of the head; all others are absorbed or reflected.

In *The Sonar of Dolphins*, published in 1993, Whitlow W.L. Au, an authority in this field, cited numerous studies to insist that the melon's function as a transmitter of sounds into the water is in doubt, because it cannot reflect them enough to act as an acoustic lens, as others suggest. Furthermore, it cannot, in itself, explain the directional sound transmission of Odontocetes. He argues that there is no direct evidence to support either theory of sound production, and that many anatomical structures could be involved in this process. Obviously, further studies are needed in order to settle the matter once and for all.

Supersonic click emissions in Odontocetes may also serve another purpose. Some researchers believe that these sounds carry enough energy to literally stun prey. Tests performed with animals in captivity seem to provide evidence of this capacity and help to explain how the relatively slow-moving sperm whale can catch and feed on prey like the fast-swimming giant squid (*Architeutis*).

The sperm whale probably sends a soundwave beam of an intensity proportionate to its size in the direction of a prey detected with its biosonar. Such a function would explain why totally blind sperm whales or those with broken lower jaws are as healthy as normal individuals—and why the squids found in their stomach are never bitten!

Mysticetes also have a vast repertoire of sounds, including some that are quite melodious. Many people have been moved by the haunting songs of the Humpback whale.

The role of sounds emitted by Mysticetes is not very well known. They are produced at relatively low frequencies and are extremely powerful. The moans of a Blue whale can exceed 188 decibels (dB), making them the most powerful sounds produced by a living creature. In comparison, a jet engine plane produces sounds that range between 140 and 170 dB!

Low-frequency sounds can also travel great distances in the ocean. Winn and Winn (1978) have recorded Humpback whale songs from a distance of 32 km, and Cummings and Thompson (1971) detected sounds from a Fin whale 160 km away. Even better, some scientists, such as Tomiline (1974), believe that Mysticetes can hear one another at a distance exceeding 1,000 km (621 mi.) through low-frequency sound signals (20 Hz) propagating in the acoustic channels of the ocean (SOFAR: Sound Fixing And Ranging). These channels form between water layers of different temperatures and therefore of different densities. Infrasounds and low-frequency sounds propagate through these channels without being absorbed, and could allow animals having a wide distribution range to remain "acoustically connected" and to synchronize their activities, much like the elephants do.

Giant squid at the Smithsonian Institution.

Can Mysticetes use their sounds as a sonar? Many people think they cannot. However, Beamish and Mitchell recorded the clicks of a Blue whale trapped in the ice, by means of hydrophones. The two scientists suggested that these clicks were used to echolocate schools of krill. Oddly enough, the clicks became more frequent when a lead weight was placed near the whale. The careful positioning of three hydrophones used to record the sounds has allowed Beamish to establish that the source of propagation was located in the anterior portion of the head. Might this be the first confirmation that certain whales can use the bones of their skull to directionally transmit ultrasounds as sonar waves? (Beamish and Mitchell, 1971; Beamish, 1974, 1979)

Mysticetes' biosonar likely gives a cruder acoustic image than the one produced by Odontocetes. But as we mentioned above, a lack in image quality would be more than compensated by the fact that baleen whales feed on prey that often gather in enormous shoals. Moreover, the much larger range at which their low-frequency sounds are emitted would allow for the detection of shoals of krill or fish from much greater distances. It may be used, too, to find the depth of the bottom. I have never seen a whale with a crushed rostrum. When huge whales dive in relatively shallow but turbid waters, as in the St. Laurence, the bottom is not that far away and must be quite invisible. They surely have a way to detect it before hitting it! It may be used to recognize some big features of the ocean floor, which would be used for navigation, or sharp differences in salinity or temperature of different masses of water, which would allow them to find productive zones in the oceans (like upwellings for example).

Pinnipeds, too, are able to produce all kinds of sounds, but we do not know much how they make them, because few studies have been devoted to them. Otariids produce barking sounds in air as well as underwater. When they bark underwater they do not emit bubbles and produce 2 kinds of sounds. It shows, that the implied mechanisms must be varied and complex. Phocids too can emit several kinds of sounds. Most of them have vocal chords and sacs near the pharynx and/or along the trachea. It is not known how they use them to produce sounds. The Bearded seal (*Erignathus barbatus*) makes very long songs ended by a moan. The posterior part of the trachea is made of a long elastic membrane. This membrane must participate in sound emissions. The Hooded seal, (*Cystophora cristata*) can inflate a sac on its head and even, by closing one of its nostrils and blowing in the other, inflate a sac made of the extension of the elastic part of the nasal septum like a red balloon. This sac is used as a visual display and to produce sounds when the seal rapidly moves its head up and down or right and left. The Walrus is one of most vocal pinnipeds. It makes varied vocalizations, from whistles to gong sounds and many kinds of grunts, either in the air or in the water. It has paryngeal sacs and the peculiar shape of its tongue can be used in sound production. The Harp seal, at least, seems to emit clicks a lot like those used by cetaceans in audio-detection (sonar). No one is completely sure they are used for the same purpose by pinnipeds. This demonstrates again how much fundamental research is needed to better understand those animals.

Fin whale.

Is the language of cetaceans able to convey abstract ideas? A good many people, their thinking influenced by the size of the cetacean brain, by the myths that surround whales, by the desire, conscious or otherwise, to find our kindred spirit in the animal world, or by an animalist philosophy, have tried to prove it. Some continue to support John Lilly's views on this subject, despite the fact that competent cetologists today challenge the results he obtained in his experiments. Wilson (1975), for example, writes: "Lilly's books are misleading to the point of bordering on irresponsibility" (in Herman and Tavolga).

One of his experiments, often repeated in a biased or incomplete manner, or without commentary, would seem to suggest so. Psychologist Jarvis Bastian working with Bottlenose dolphins in the U.S. Navy laboratories at Point Mugu placed two dolphins in an aquarium. He trained a male named Buzz to mimic a female named Doris when the latter pressed on a series of plates in a particular sequence in order to activate a light. An opaque but acoustically transparent barrier was then added to separate the two animals and the test was repeated. The two dolphins exchanged a series of whistles, leading Bastian to believe that Doris had communicated the correct sequence to Buzz, who continued to imitate the female perfectly. When Bastian then added an acoustically opaque barrier, Buzz made mistakes. Conclusion: Doris communicated to Buzz abstract information; hence, dolphins do have a language! Unfortunately, it is not that simple! Bastian invalidated his own conclusion by showing that Buzz was able to determine Doris's position by sonar, despite the presence of the acoustic barrier. Furthermore, the sounds that Doris emitted were always the same, regardless of the action Buzz was supposed to mimic (in Tomiline, 1974).

Two Bottlenose dolphins (*Tursiops truncatus*).   Michael W. Newcomer

People have also tried to prove the existence of language, at least in Odontocetes, by comparing its whistles to the whistling language of shepherds in the Canary Islands or Basque regions. Here again, we must practise caution. Whistling languages *translate* a spoken language, with its vocabulary and syntax. They must be learned as any other language is, over time. From birth to the age of adulthood, whales acquire very few sounds—from 6 to 20 (the Bottlenose dolphin acquires 30). By contrast, the typical human being will have learned up to 5,000 words by the age of six.

Yet another hypothesis suggests that from the moment Odontocetes can make an acoustic image of their world from the echoes they receive by means of sonar, they could also communicate these echo-images to their conspecifics. The receiving whale simply has to decode such a message as if it were its own! Once again, it does not seem likely that a whale can differentiate between normal echoes that might come from its foraging pod members (which it must ignore, if it wants to recognize its own), and such an "echoconversation!" We can only be impressed by the perseverance shown by some people's efforts, at all costs, to humanize cetaceans and by the dimensions such myths concerning whales can take.

Blue and White dolphins. *Dmitri Marguérat*

The term "dialect," used to illustrate the fact that individuals and family groups produce "signature sounds" which are used by scientists to differentiate individuals and family groups, contribute to the perseverence of this misunderstanding. In this case the term "dialects" has not the same significance than when used to connote the local variations of a spoken language allowing people to pass abstractions. There, we have only variations in the pitch and frequency of the emitted sounds, a little like the different tones in human voices which allow us to recognize particular individuals.

Dolphins on the move. *Michael W. Newcomer*

All serious studies to date suggest that the language of cetaceans cannot convey abstract notions, but rather concrete information, as determined by factors within their immediate surroundings and made accessible through their senses.

Naturally, questions concerning cetacean intelligence do come to mind. But in order to discuss intelligence, we must first define it. If, on one hand, we mean by intelligence the faculty of living things to resolve problems related to their survival through appropriate mental processes, then whales are certainly intelligent, just as dogs, cats, and flies, and other living creatures are. Whales would certainly not have survived 18 million years if it were otherwise (the test is several hundreds of millions of years for flies!).

On the other hand, if we define intelligence as a faculty whose scope extends beyond survival skills to include solving abstract problems, reasoning abstractly, imagining, distinguishing oneself from the environment, anticipating, creating, communicating by means of a continually evolving and complex language—all characteristics attributable to human intelligence—then whales seem to lack these qualities and are no different from other animals. How could we otherwise explain behavior patterns such as the annihilation of a herd during mass strandings? Or why do dolphins drown in nets used in purse-seine fishing, when they could easily have jumped over them? Or why do dolphins that have just been caught or have just lost a partner "commit suicide," that is, repeatedly hurl themselves against the sides of their basin until they finally die? How otherwise can we explain the fact that whales are regularly caught by hunters, despite these animals' long-range echolocating capability, or the fact that these pods contain individuals who have had previous, painful encounters with humans (as seen by the scars they bear) but are nonetheless still caught?

Convoluted folds (gyri) of a cetacean brain.

Many people are puzzled by the whale's large-sized brain and its highly convoluted folds. We tend to associate a large brain with higher intelligence (according to the second definition, above). We also often associate intelligence with the development of the cerebral cortex. A high degree of folding of the cerebral cortex suggests a large cerebral cortex and, therefore, superior intelligence. Yet the brain of cetaceans is not that large, in proportion to their body size. Furthermore, brain weight, even when considered with respect to the total body weight, does not mean very much. In *Homo sapiens*, it represents 1/34 of the body weight; in the black spider monkey (*Ateles paniscus*), a South American monkey, it represents 1/15 of its weight (Tomiline 1974). Can we then conclude that the spider monkey, a relatively primitive primate, is more intelligent than human beings? It is not brain size but the number of brain cells, as well as the number of connections among them, that indicates brainpower. In other words, the structure is more important than the volume. The organization of the neurons within the cetacean brain is somewhat similar to that found in mammalian insectivores, whose cerebral organization is quite primitive.

As for the cerebral cortex, other faculties beside intellectual ones are involved in that part of the brain. Much of it serves to collect sensory information. As we have already seen, cetaceans rely mostly on the analysis of auditory information to create mental representations of their world. The auditory centers are therefore important and occupy a major part of their brain. In addition, the cetacean body is generally large; such large surfaces must be able to react immediately to changes in water pressure to avoid creating turbulence while swimming. These functions presuppose the presence of numerous sensory receptors and accompanying neurons in the brain centers to permit all necessary and appropriate actions. With this in mind, the reason for the cerebral cortex's large surface becomes clear.

Brain endocasts from the craniums of a human, a Harbour porpoise, and the Beluga.

Cetaceans have evolved in a world so alien to ours that, even if they had mental abilities similar to ours, it is doubtful that we could communicate with them in any worthwhile way. Their vast cortex allows them to carry out mental processes wholly unlike ours. It is important that our attitude towards whales be free of the smallest hint of anthropomorphism. They are fascinating animals and do not need to possess human characteristics in order for us to appreciate them more. Let's enjoy whales for what they are, which is more than sufficient!

Pinnipeds brains are of a size very much like what one can find in terrestrial advanced carnivores. The cerebellum is bigger relative to those of land carnivores (as in cetaceans, too). It must be in relation to the necessity to coordinate their movements in the aquatic part of their life, where gravity is playing a lesser role and they can move in all 3 dimensions. Olfactic lobes are present, but are less important than in the brains of some land carnivores of same size.

Ursula Tscherter

STRANDINGS

Strandings, or beachings, illustrate one type of whale behavior that may seem unintelligent or maladjusted. Throughout the world, a great number of cetaceans are washed ashore each year. Many of these cases involve dead animals that end up on the shore, as would any floating debris. In other cases, however, living animals will strand and later die.

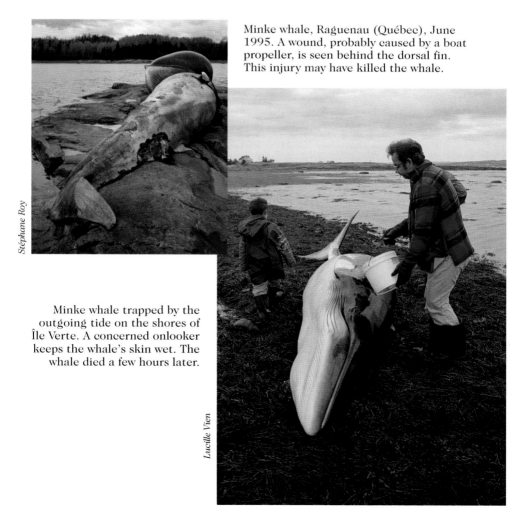

Minke whale, Raguenau (Québec), June 1995. A wound, probably caused by a boat propeller, is seen behind the dorsal fin. This injury may have killed the whale.

Stéphane Roy

Minke whale trapped by the outgoing tide on the shores of Île Verte. A concerned onlooker keeps the whale's skin wet. The whale died a few hours later.

Lucille Vien

How can we explain this type of behavior? Aristotle, who was interested in almost everything, admitted he could not. Since his time, many people have tried; it would have been better if some had done as the great philosopher and naturalist did, and foregone the effort to explain.

Almost every possible reason has been put forward to explain strandings, from meteorological conditions to terrestrial magnetism, from naval maneuvers to pollution, illnesses, and parasites, and, why not, the weariness of living in a world that mankind's folly has made unbearable.

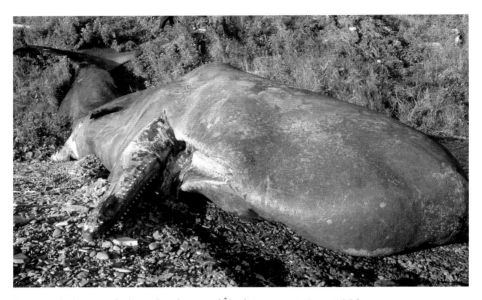

Sperm whale stranded on the shores of Île d'Anticosti, June 1992.

Suicide, an act performed by humans, either from lassitude, discouragement, or the inability to face what seems to be an inextricable situation, must be excluded as a possible explanation for the stranding of cetaceans. Some behaviors that apparently indicate suicide in animals are often cited by animal rights activists to defend the irrational extremism of their views regarding mankind's abuse of animals. For example, Tomiline refers to a few cases in which dolphins have died by hurling themselves against the sides of their aquarium. Can we conclude that the sadness at losing a partner or freedom has led them to this dire extremity? Is it not more logical to conclude that we have simply witnessed a thoughtless and stereotypical escape behavior in which the walls of the aquarium represented a barrier never recognized as impassable? An animal may adopt a behavior that is eventually fatal, but death is certainly not what it is seeking. Instead, the animal flees an unpleasant situation for which no interpretative grid exists, merely a series of repetitive gestures related to innate behaviors which causes its demise.

A closer look at different strandings reveals that they are not all alike and that some are easier to explain than others. In 1983, on the shores of Cap Tourmente, near Québec City, a young Fin whale measuring 13 m (43 ft) washed ashore alive. It was most likely trapped by the strong tidal currents characteristic of the area. Similar circumstances are probably behind the strandings some ten years later of a Minke whale at île aux Coudres, of another Minke whale in the estuary of Bersimis River, and of two Northern Bottlenose whales in the Montmagny region, east of Québec City.

Northern Bottlenose whale stranded near Montmagny, Québec.

Early one morning in October 1996, a young Minke whale was washed ashore by the outgoing tide on île Verte and died some 16 hours later. The tides may not have been the only factor in this case or the ones mentioned above. The fact that some of these whales were far from their usual habitat (especially the Northern Bottlenose whales which are almost exclusively PELAGIC) and that they ended up in freshwater could also have played a role.

A Blue whale stranded close to Sept-Îles attracts a crowd of the curious.

Witnesses to the strandings of Minke whales have reported that these whales started to struggle only when the water was low enough for its weight to become a problem. Could their extremely thin and sensitive skin cause them to stop moving the instant they make contact with the bottom, preventing them from making a timely escape? Or, because most strandings among Mysticetes involve juveniles, could the cause be attributed then to inexperience? It is unknown whether whales that strand and manage to escape have learned enough from the adventure to prevent its recurring. We must also bear in mind that, dramatic as these strandings may seem, they are definitely not common occurrences, given the fact that hundreds of thousands of cetaceans do not strand.

Other cases of single strandings have involved older or diseased animals. Are these whales "conscious" of the danger of drowning? Do they try to avoid drowning by voluntarily beaching? The New Zealander Frank Robson, who has studied hundreds of strandings, is convinced that this may be so. Concerning the stranding of sick or older whales, he has observed also that it is futile to try to push the whale back into the water because it will keep beaching, until it dies.

A female Beluga whale stranded on Île Verte in May 2000, probably injured by a boat propeller. Its skeleton is exhibited at the Skeleton Museum on the island.

Mass strandings may result from a herding reaction brought on by a distress signal sent by one or more members of the group that have beached, for reasons we will discuss later. Stranded whales emit agonizing distress signals. We have seen that whale sounds have a social function, and that distress signals will usually bring about herding behaviors. In responding to the call for help, the rest of the group will eventually beach. Numerous well-documented observations seem to confirm this scenario (Tomiline).

Mass stranding of Pilot whales.

Frank Robson has observed that if one kills the first whales of a group to strand, the social bond with the rest of the pod is broken. The others will not beach. This method may seem cruel, but it eliminates the useless suffering of the stranded whales—a stranded whale can live for 48 hours before dying—and it saves the rest of the herd. Although doing so may not be effective for all cases of strandings, and determining which animal to kill to prevent additional beachings is difficult, this method is worthy of further investigation, especially for larger whales. Perhaps we can then avoid situations similar to those involving two herds of Sperm whales that died on the shores of New Zealand—59 females and juveniles in 1970, and 72 females, juveniles, and the dominant bull in 1974—or those cases of hundreds of Pilot whales that regularly strand on the shores of Newfoundland. We must also remember that when a large-size whale washes ashore, it will literally be crushed by its own weight. Breathing becomes extremely difficult because the animal's respiratory muscles are incapable of sufficiently dilating the thoracic cage to ensure adequate pulmonary ventilation. Furthermore, the blood flow in the part of the body that rests on the ground is cut off by the animal's weight. Tissues begin to die from the prolonged absence of a steady blood supply, much like when a tourniquet is applied too long. If the whale is returned to the water, the gangrenous tissues cause extensive infection that, most often, leads to the animal's death.

Let's return now to the reasons—for there is probably no single one—why solitary animals or the first individuals of a herd beach. Among these explanations we may include pathologies that affect the auditory system of whales and thus their ability to echolocate, such as infections of the inner ear or its invasion by parasitic NEMATODES. Some biopsies have revealed the presence, sometimes in vast numbers, of such parasites. What remains unclear is whether or not they represent a nuisance to the host or if indeed so, to what extent.

Other factors that may explain stranding include: shallow coastal areas with long, gentle slopes; storms in which air and sand mix with the water to make echolocation ineffective; or intense hunting during which animals inadvertently beach while chasing a prey with a smaller draft. These factors cannot explain, however, strandings that occur on rocky, abrupt coastlines or during calm periods.

Petrotympanic bone of a Harbour porpoise. Parasitic nematodes are clearly visible.

A new explanation has recently been gaining popularity among specialists. A great number of strandings (if not all of them) may well be caused by navigational errors brought about by disturbances to the earth's magnetic field.

It is important not to confuse *orientation*—the ability to ascertain and maintain a direction—with *navigation*—the ability to ascertain the position of a point in space and to know how to get to it. For instance, an individual with a compass, whether in a boat or on a plane, will maintain a constant direction but may very well miss an intended target if an air or water current causes him to deviate laterally. On the other hand, a person equipped with a map (real or mental) and capable of recognizing certain topographical elements or relief features can reach his desired destination simply by following a path, without the use of a compass, however tortuous that may be.

Stranded herd of Pilot whales.   *N.E. Aquarium*

Margaret Klinowska (1985) of the United Kingdom has suggested that cetaceans use the total geomagnetic field as a navigational map and a system based on its regular fluctuations as a means to measure time. In this way, they can establish their relative position on that map, assuming they can also estimate their speed. Many organisms, such as pigeons, bees, and other migratory animals, including human beings, use similar methods for navigation. (This might explain why one of my orienteering students never got lost in the forest. He had told me, and I didn't believe him at first, that he never needed a compass to find his way out of a forest, and… he was right!)

The geomagnetic field is formed principally by electrical currents—similar to those produced by a bar magnet, with its north and south poles—that flow through the metallic, fused mass of the earth's core. The ion flow in the jet streams of the upper atmosphere is also part of the field, and is subject to regular daily fluctuations. The geomagnetic field can also be affected by solar activity, or what we call magnetic storms. These storms will completely change the normal rhythm of the daily fluctuations. Finally, the shape of the earth's magnetic field is not the same everywhere: it is influenced by the subterranean geology and metallic ore deposits in the underground. The magnetic topography is analogous to a set of hills and valleys; it is weak or strong according to its peaks and lows. This field could thus be represented as a contour map, complete with its lines of elevation, or isopleths, much like a relief map of the earth's surface. When travelling, whales seem to follow magnetic contour lines, one higher to the left and one lower to the right, for example, or vice versa.

Stranded Common dolphin, still alive, on the shores of Mauritania.

How can whales follow contour lines? They might do so by means of a metalloid substance of iron oxide crystals (magnetite), found among other metallic substances, in or around their brains (J. Zoeger 1981). These crystals are found in areas that include the parts of the brain that regulate the whale's biological clock and integrate sensory information. Other structures are probably used to detect the earth's magnetic field and its use in navigation, but these remain to be discovered or ascertained.

Migrating whales thus simply need to be able to read the "magnetic map" of the area they have to cover, as well as evaluate their speed and elapsed time. In other words, their "magnetic sense" is not used as a compass to orient themselves, but in a manner similar to our taking into account topographical details such as hills, valleys, and other landmarks that we have memorized, to aid us in finding our way. Whales also gain a notion of elapsed time by adjusting their internal clock according to the fluctuations of the geomagnetic field. Each day, they reposition themselves on their map in accordance to the regular maximum and minimal fluctuations in the field's intensity. How do whales estimate their speed? Perhaps tactile corpuscles on their lips and melon, or their vibrissa detect pressure variations resulting from the constantly changing speed of the water around them.

Disturbances caused by magnetic storms can disrupt the means whereby adjustments to normal variations are used to reset their clock. In this case, a whale would lose or gain time. Consequently losing its position on the map, it would follow the topography that corresponded to another day in its migratory route.

Magnetic contour lines do not always run parallel to the coast; sometimes they cut across it at right angles. By following its mother, a calf has learned to avoid these traps and as long as it can accurately locate itself, it will continue to do so. Unfortunately, a badly-set internal clock could shift the animal's positioning with respect to the correct one (i.e., the one on its memorized map). As a result, although its mental map is telling the animal that it is travelling along the shoreline, it will head towards it and beach. The rest of the herd will do likewise. Here is perhaps the reason whales systematically return to the shore, even when they are pushed back into the water. The only usable guide that they have is the contour line of the magnetic field they are following at that moment.

- ● Coast Guard Station
- ▪ Strandings of Dead Animals
- ◀ Strandings of Live Animals
- ▬ Contour Lines of the Magnetic Field. Isopleth intervals of 10 nanotesla

Map according to Klinowska. We see that the strandings of live animals occur where isopleths are perpendicular to the shore.

Klinowska, who has closely studied whale strandings on the English coastal areas, has observed that whenever carcasses of dead animals land on shore, they do so where the currents are favorable. On the other hand, live strandings occur exclusively in coastal areas characterized by magnetic anomalies (field lower than average, for example), in "valleys" that are perpendicular to the shore or blocked by islands, and after magnetic storms. She also points out that pelagic species strand proportionately more often than coastal ones. The reason seems quite logical. When near the shore, pelagic species find themselves in a relatively unknown territory and rely on the magnetic topography for navigation, whereas coastal species, accustomed to a familiar landscape, depend mostly on the geographical topography of the area. The work of Kirswink, Dijon, and Westphal (1985) on strandings along the East Coast of the United States seems to confirm these findings.

This navigational system is quite simple. It requires only a map and a clock. It is unaffected by the drift of the magnetic poles, even their reversal, since it is based on the magnetic topography, which is determined by the geology of the area. It is a very efficient system given that strandings are very rare, as we noticed earlier. Cetaceans can probably "realign" their navigational system by the use of their sonar, which gives them a general idea of the geographical topography of the bottom, or by resetting their internal clock, which allows them to correct for navigational errors.

It is becoming more and more reasonable to view strandings as the result of navigational error. Although there may be other causes, the much higher frequency of strandings *in* areas of magnetic anomalies or *following* major disturbances to the normal fluctuations of the geomagnetic field probably makes this error the principal cause.

I have witnessed the stranding, then the death, of 7 White-sided dolphins (*Lagenorhynchus acutus*) on the shores of Île-Verte, in September 1999. In spite of all the efforts a good samaritan from the island and I deployed (it meant spending 3 hours in very cold water!), they all died on the shore where they were returning no matter what we did. I think that they swam in the estuary, following a bearing SW, arriving in a region they did not know (we rarely see this species in the vicinity of Île-Verte) When it was time for them to go back they followed the opposite bearing (NE: they had memorized the ingoing bearing). But maybe they were at the time too near the south shore so their course took them to the island. If they had hit the island almost at the tip, they could have, by trial and error (as do all animals in this situation), found their way out of this trap. Sadly they hit the coast too far from the tip of the island. To verify this hypothesis they should have been transported on the other side of the island, but we were not able to constrain them and the one we tried to transport hurt itself so much that we had to put it back in the water. Anyone possessing the same kind of intelligence as humans would have been able to reason this problem and follow the coast until the way out was found.

Four of the seven White-sided dolphins that stranded on Île Verte, probably the result of a navigational error.

Fin whale at Pabos, July 2000

Their form of intelligence did not allow them to do that and, as they were not able to control the situation, they all died. It does not make them stupid animals, only beings whose brain is not made to solve such problems other than by trial and error. Evidently in this case that was not enough. Maybe if they had survived they would have memorized the event and more easily solved another similar problem. But they could not have transmitted their experience vocally to other dolphins met later; their language does not allow transmission of such abstract notions.

Fin whale drifting close to Tadoussac.

Ursula Tsherter

The same whale at Point-au-Père.

The multiple fractures to the humerus may have been caused by a violent collision with a boat.

At the end of 2004, a greater number of strandings than usual occurred on the coasts of Australia and New Zealand. Scientists from the Marine Biology department of Tasmania University have discovered that it is a cyclic phenomenon, related to the force of the south and west winds which peaks every 12 years. Those winds would push the colder currents nearer the coasts, and as we will see in the next chapter, their waters contain large amount of food. Following them, the cetaceans would come too near the coasts and some would eventually strand. As we have seen in gregarious species (like Pilot whales), only one specimen making an error and stranding can push the whole group on shore.

NUTRITION

Like many other animals, marine mammals have, in the course of their evolution, come to occupy the ECOLOGICAL NICHES left vacant by the extinction of the large marine reptiles. They have benefited from the vast richness of the marine environment and shared these resources by developing both efficient and varied methods of capturing a vast array of prey.

We must once again discuss Odontocetes, Mysticetes and pinnipeds separately since they use very different feeding techniques.

ODONTOCETES

Most Odontocetes share an anatomical characteristic that leaves no doubt as to their predatory way of life: PREHENSILE teeth. Well-developed and all alike (an example of homodont dentition), these teeth have given toothed whales the ability to grasp and even tear off large chunks of flesh from moving prey. Since this type of dentition makes mastication ineffective, prey, or parts of them, are swallowed without being chewed. Some whales, such as *Ziphiidae* (beaked whales) have but one pair of functioning teeth. In females, these teeth almost never erupt but remain hidden in the gums throughout their life. Teeth in Odontocetes probably serve more as weapons during violent combats over females during the rutting season than for feeding. The Sperm whale has up to 40 teeth in its lower jaw (mandible), and each one can weigh more than 500 g (1.1 lbs.), an impressive row of teeth, indeed!

Mandible of a great sperm whale, probably a male. *New Bedford Whaling Museum*

Nevertheless, it is unlikely that Sperm whales use these teeth much for feeding. They erupt at around the tenth year of life, two years after weaning. Obviously, these whales cannot have waited for the appearance of their teeth to begin feeding. We have also seen old Sperm whales whose mandibles were broken, probably by a harpoon or a fight, and which subsequently healed at a right angle. These animals were no thinner than other Sperm whales with functional mandibles.

Maxillary teeth (those in the upper jaw) are vestigial and fall out early, since they are not implanted in the bone, but only in the gums.

These animals probably feed by swimming with their mouth open, and letting their mandibles trail along the ocean floor. They capture their prey through suction feeding, possibly after having stunned them with ultrasound. Drowning is avoided because their larynx terminates in the shape of a closed beak to keep water out.

Distorted mandible of a sperm whale, *New Bedford Whaling Museum.*

Deformed mandible. *Galerie d'anatomie comparée, Paris*

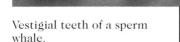

Vestigial teeth of a sperm whale.

Longitudinal Section of the Head of an Odontocetes

Larynx of a Beluga. The orifice of the internal nostrils is to the left.

Half-open larynx of a Beluga.

The larynx is set deep into the orifice of the internal nostrils or choanae. It is held firmly in place by a powerful muscle, the palato-pharyngeal sphincter, at the point where the beak emerges. The larynx cannot come free of the internal nostrils without the voluntary relaxation of the sphincter. Swallowed prey, no matter what its size, must travel either to the right or the left of the larynx, more likely to the right owing to the skull's asymmetrical shape. The animal cannot breathe through its mouth so food cannot enter the lungs. Suction is achieved by an up and down movement of the tongue and of the HYOID APPARATUS. Some toothless species, like the beaked whales, have 2 V-shaped grooves under the throat. Those grooves must allow the pharyngeal region to expand more during suction. The slippery species usually eaten (squids) are sucked up more deeply in the mouth and so are more easily kept in.

By contrast, the Killer whale most definitely uses its teeth, as seen by the scars it has left on prey that managed to escape, such as the large pinnipeds or other cetaceans, and by the numerous documents where they are seen pulling pieces from their prey.

The V-shaped grooves under the throat are peculiar to beaked whales.

Suction feeding in the Beluga: 1) the mouth opens; 2 and 3) the mouth is in the shape of a funnel; the tongue is visible and elevated 4) the tongue lowers itself in order to create suction; it is no longer visible.

Whales have a multicompartmental stomach, somewhat similar to that of their closest living relatives, the ruminants, with whom they share a common ancestor.

Scars on the tail of a Humpback whale left by the teeth of Killer whales.

Opened stomach of a Beluga. The 3 compartments are differently colored.

The first compartment serves in general to accumulate food and prepare it for the action of gastric juices. By examining stomach contents, we can get a good idea of the animal's diet.

Odontocetes have intestines generally 15 times their body length. I have verified that in White-sided dolphins and in the Sperm whale. A juvenile specimen, stranded in Anse à Valleau (Québec) measured 13 m and had intestines 184 m long. We do not know why they are so long, since their prey does not seem to be that difficult to digest. It is maybe because, during deep diving, the digestive system does not receive much blood. Its efficiency must be at its maximum when the animals are near the surface.

Odontocetes capture their prey in many different ways. Some hunt alone and locate their prey by sonar, which they capture one by one. Others, such as the Sperm whale, wait in ambush at a fixed depth and use their sonar to find food. It is also possible that they stun their prey by ultrasound. Finally, others form groups to hunt, demonstrating a certain degree of co-operation.

Beluga pancreas.

The digestive system of a White-sided dolphin. As in the Sperm whale, the intestines are 15 times the animal's body length.

Measuring the intestines of a Sperm whale (more than 184 m), searching, in vain, to find ambergris. Upper right, a piece of ambergris, *New Bedford Whaling Museum.*

Cetaceans that form large groups to forage, notably dolphins, move as a unit that stretches out farther in width than in depth. Pilot whales, whose groups sometimes number in the hundreds, form the widest front when they forage. They can cover a range of more than two kilometres at a time. When food items, for example, schools of fish or squid, are located, they move into one of two formations.

Dispersed formation—the group forms subunits, in which each individual displays its own behavior. A certain degree of synchronicity in the behavior of the group is maintained, however. For this to happen, some sort of communication among the group members must take place.

Co-operative formation (coordinated hunting)—individuals seem to work together to keep their prey from dispersing or to force it into an area where capturing is easier. This type of behavior is seen in some dolphins and in Killer whales.

Pod of orcas in the Mingan Region.

Videos shot near Norway show Killer whales trapping their prey, herring, by surrounding them, all the while emitting sounds. Then, one by one, they violently strike at the school of fish with their flukes and feed on their dazed victims. The entire process is carried out by means of a distinct and probably hierarchic sequence. The herders continue their work while the "hitters" collect their prey.

Belugas in the St. Lawrence River, near île Verte.

Opposite Page:
Fascine fishing, île Verte, Québec.

I have seen the Beluga demonstrate a similar behavior pattern. On three occasions near île Verte, in the St. Lawrence River, I observed groups of five or six Belugas form circles of about fifteen meters (49 feet) in diameter. Each time, one would periodically leave the circle. My feeling was that it did so to catch one or many prey items herded by the group. The individual whale must have done so while diving, since I would see it surface in the middle of the circle and rejoin the formation afterwards. All three times, the activity lasted between 15 and 30 minutes. In one of those cases (1974), I managed to get my motorboat to within ten meters of the whales. Normally, they would have been unwilling to be approached, as hunting had stopped only two years before. At that close distance, and with my motor turned off, I drifted beside the pod for some time and observed its activity. Herring was regularly caught in the fascine fishing gear used by the island residents at the time, so perhaps it was herring they were feeding on. I cannot say for sure, but something did indeed take place that clearly required a coordinated effort.

I observed similar behavior by a group of Bottlenose dolphins (*Tursiops truncatus*) in Mauritania, but this time, one of the dolphins was rapidly moving at the surface of the water while the others seemed involved in herding prey. This behavior lasted about an hour and their activities appeared coordinated. The members of the group, between five and ten, covered an area of several dozen metres between successive "feeding" operations, but always remained grouped.

What do Odontocetes feed on? Their diet consists of PELAGIC, DEMERSAL or BENTHIC fish, CEPHALOPODS, crustaceans (pelagic or benthic shrimp), worms, molluscs, and even mammals and birds. Some species, such as the beaked whales, are almost exclusively teuthophagous (Gr., *teuthis*: squid).

Others have been described as excessively voracious as Slijper's famous drawing of a Killer whale with the contents of its last meal clearly suggests. Slijper tells us in *Whales and Dolphins* that the eminent cetologist Eschricht found, in the foregut of a 7.5 m (24.6 ft) Killer whale, 13 perfectly intact porpoises and 14 seals. An additional seal had supposedly been found lodged in the whale's throat. This poor orca, which could not have weighed more than five tons, would have had close to two tons of food in its stomach, definitely enough for fatal indigestion!

Last meal of Slijper's Killer whale.

The real story dates back to 1866 when the highly respected Eschricht (for whom a family of whales, the Eschrichtiidae, and one species, the California grey whale, *Eschrichtius robustus*, were named) examined the stomach contents of that famous whale. He had, in fact, found the remains of 13 porpoises and 14 seals, but he had taken the precaution, however, to point out that the remains were highly fragmentary and that some were minute. And that is how fish stories are written!

Silhouettes of an eared seal and a surfer, seen from below.

Nonetheless, the Killer whale is, all things considered, a formidable predator, able to feed on almost anything that swims, including Mysticetes. Oddly enough and for unknown reasons, it does not seem interested in having humans on its menu. Indeed, no well-documented evidence has been presented to show that they have eaten a human being. The only known case is that of an attack on a surfer who, when seen from below the surface, must have looked very much like the sea lions commonly hunted by those whales. The assault ended as soon as the whale "realized" its mistake.

The "resident" Killer whales of British Columbia feed almost exclusively on salmon. The "nomads" feed mostly on seals and other mammals.

MYSTICETES

When we examine the skull of a Mysticete, we immediately see that it is missing teeth and alveoli (teeth sockets) in both the maxillaries and mandibles.

Skull of a Fin whale.

117

The maxillaries are more or less flattened depending on the species and serve to support a structure admirably well adapted to the way in which Mysticetes feed: the baleen (whalebone). Mysticetes do not feed by capturing prey items one at the time the way Odontocetes do. They do so by eating vast quantities of food items that they filter out from considerable volumes of seawater.

Baleen plates have an epidermic origin, just as our hair and nails do. Their evolutionary origin, however, remains a mystery, one that creationists often use to discredit evolutionary theory. But a close look at the palate of an animal reveals the presence of transversal horny ridges. It is possible that baleen plates have developed out of similar ridges (see the chapter on paleontology). Both cases are examples of secondary epidermal formations.

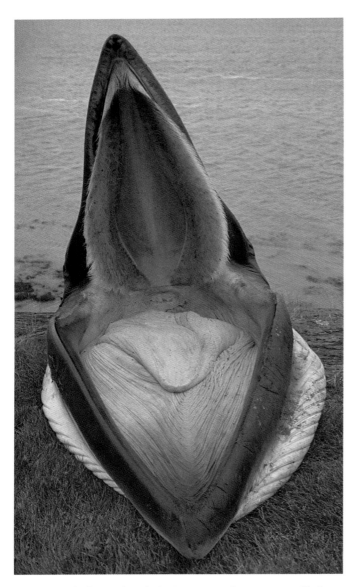

Head of a Minke whale. In its mouth, two rows of baleen plates are seen, île Verte, Québec (October 1996).

A baleen plate is formed by two horny plates separated by a series of coarse and fine filaments or bristles (similar to gluing two fingernails together with the whiskers of a cat between them).

The bristles and horny plates grow continuously throughout the life of the whale, but the bristles wear away much more slowly than the DISTAL extremity of the plates. As a result, the outer plates bear a fringe of bristles that extends many centimetres beyond the horny matrix.

Horny ridges on the palate of a moose.

Roots of the hair of a Minke's baleen plates Baie Comeau.

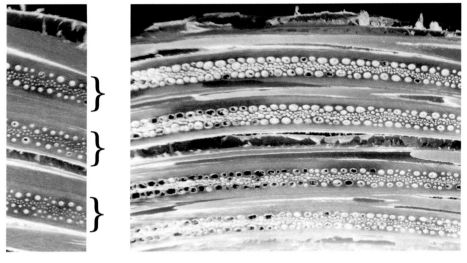

Cross-section of a Fin whale's baleen plates. Notice the bristles in the horny matrix.

Baleen of a Minke whale seen from the exterior.
We see the lip in the upper part of the photo.

Mysticetes have from 200 to 450 baleen plates on each side of the mouth. A distance of 5 to 10 mm (0.2 to 0.4 in.) separates adjacent baleen plates. The smallest plates are at the extremities whereas the longest are at the center.

The bristles of the plates' inner edge overlap to form the filter bed that allows water to flow through freely but efficiently retains prey items.

Minke whale's baleen plates with their overlapping bristles, seen from inside.

The smallest PLANKTONIC elements are the least affected by the sieving process. Indeed, if the filter mat of baleen fibres were any denser or the gaps between the plates any smaller, the flow of water out the mouth would be slowed down. Establishing a correlation between the fineness or the number of bristles and the size of prey taken is difficult, even though the desire to do so is logical enough. But no significant differences exist between the baleen fringes of a Fin whale, which feeds on fish, and those of a Blue whale, which feeds almost exclusively on krill and copepods. It seems that the baleen's evolutionary priority was to establish a minimum threshold of resistance to the flow of water.

Furthermore, fringe size changes with age, but whether or not prey size also varies remains to be determined.

No observable difference is seen between the size of the bristles of a Blue whale (above) and those of a Fin whale (below).

This feeding system is extraordinarily efficient, allowing a Blue whale, for example, to capture the four tons of tiny "shrimp" it needs daily, each weighing about one gram.

We will look at the disastrous effects of pollution on cetaceans later. It seems appropriate here, though, to point out the possible consequences of an apparently insignificant gesture, the disposal of plastic bags in the ocean. We can imagine what happens when one of those bags ends up in the baleen plates of a whale. Then, fellow readers, who will be our Jonah to go free the whale of a hindrance that could greatly reduce its ability to filter water and feed itself?

It is quite possible that the turbulence created by a rapid intake of water will clear the baleen, or that the whale will manage to remove the bag with its tongue. But would it be able to spit it out? And, if the whale swallowed the bag, what would be the effect on the digestive system?

Sea turtles do not distinguish plastic bags from the jellyfish on which they feed, a much too often fatal mistake. It is already too high price to pay for human negligence. Must we add Mysticetes to the toll?

There are two types of Mysticetes: the "skimmers" and the "swallowers" or "gulpers." The skimmers are those that filter plankton-containing water by swimming slowly through the water with their mouth slightly opened. Water flows out over their enormous lips after having

Fabrice Kerleau and Gérard Mauger looking at the body of the young Minke whale killed by ingestion of plastic bags. It was probably still nursing.

The plastic bags as they appeared in the stomach of the Minke (below) and the same bags once cleaned (left).

passed through the mat of their baleen plates. Right whales and Bowhead whales belong to this group. It is apparently possible to hear the rattling of the baleen plates of a Bowhead whale (or a Greenland Right whale) skimming the surface of the water.

Right whales have a very distinctive skull. The rostrum is extremely thin and strongly curved. Its baleen plates are extremely long and narrow, reaching lengths of up to 4.5 m (14.8 ft) in the Bowhead whale and 2.5 m (8.2 ft) in the Black Right whale (or the Northern Right whale). The hairs are very thin, which can be the reason why it has been thought that their thickness is proportional to the size of prey. Right whales eat mainly krill and copepods. They were the first to be hunted, and therefore the first to be studied.

Black Right whale feeding. *Williamson*

Due to the low position of the mandible, the lips are very thick and reach up high. The throat and belly are smooth. The tongue is firm and bulky.

Head of a Black Right whale. Galerie d'anatomie comparée, Paris.

The mandible articulates with the skull in the usual manner, that is, by means of two glenoid (socket) cavities filled with SYNOVIA and separated by a fibrocartilaginous MENISCUS.

Many observers have seen these whales come and go, partially or completely submerged, often in mega-swarms of small animalcules. When they dive to feed, they can stay submerged from four to ten minutes.

As for the other Mysticetes, the "swallowers" (or "gulpers"), their skull is flatter, the maxillaries are broader, and the baleen plates are shorter. The baleen measures up to a maximum of one metre (3.3 ft) in the Blue whale, 90 cm (2.9 ft) in the Fin whale, and 30 cm (11.8 in.) in the Minke whale. Conspicuous pleats, or grooves, on their lower throat and part of their belly characterize these animals. It is from these grooves that they get their name rorqual (from the Norwegian *rørhval*; *rør*: pleats and *hval*: whale).

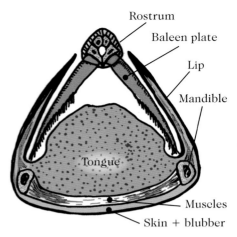

Cross-section of the Head of a Right Whale.

Cross-section of the Head of a Rorqual

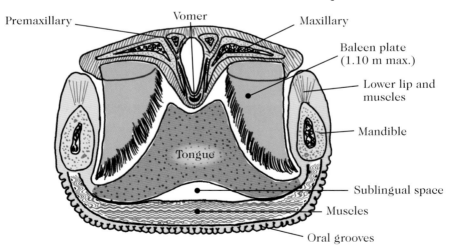

Their tongue is flaccid and the floor of their mouth can stretch considerably. Fin whales, instead of swimming slowly with their mouth slightly opened in a shoal of prey as skimmers do, use a very different feeding technique.

After having located a swarm of krill or herded a school of fish, these more rapid whales rush on their food, mouths opened wide.

Stranded Minke whale on the shores of île Verte. The grooves on the throat and belly are clearly visible (October 1996).

Head of a Minke whale. The very extensible tongue and floor of the mouth are easily visible. The esophageal opening is kept shut by the prepharyngeal sphincter located dorsally between the two rows of baleen plates, above the tongue (close-up).

The pressure exerted by the water that the whale takes into its mouth causes the elastic tissues of the mouth floor and the tongue to be pushed back in an enormous pouch (in a space meant for that, between the skin and the body: the Cavum Ventrale). This pouch, while forming, pushes on the pleats of the throat which expand in turn, a little like the pleats in an accordion. In the huge Blue whales, this pouch can contain up to 60 to 70 m^3 (15846 to 18489 gal.) of water, the equivalent of a large backyard swimming pool.

125

Head of a Minke whale. The mouth is opened, the tongue is retracting and the floor of the mouth is starting to stretch.

Once its mouth closes, the contraction of powerful muscles around the pleats of the throat and those of the tongue, which here acts as a type of piston, forces water out through the spaces between the baleen plates in a very short time.

Lesser rorqual spewing the water contained in its ventral pouch. Notice the very high pressure at which the water is expelled.

The extremity of the rorquals' larynx is not as specialized as in the Odontocetes and is not held as firmly in the internal nostrils.

Extremity of a Minke whale's larynx.

This arrangement may seem absurd when we consider the pressure at which water enters the mouth. But the presence of a prepharyngeal sphincter made by the palatoglossus separating the buccal cavity from the pharynx prevents water from entering it.

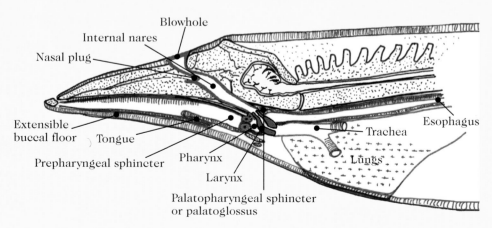

Lateral view of the head of a swallower (or a gulper) while swimming.

Thus, in spite of the enormous pressure exerted by the water entering its mouth, the larynx cannot be pulled from the opening of the internal nostrils (choanae) where it is loosely held in place by the palatoglossus muscle. The strength of that muscle and the shape of the larynx work together in maintaining the larynx within the choanae during the act of swallowing.

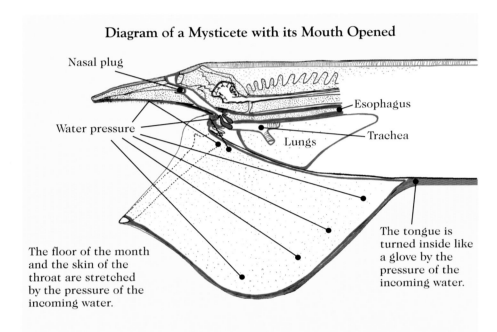

Diagram of a Mysticete with its Mouth Opened

Nasal plug
Esophagus
Water pressure
Trachea
Lungs
The tongue is turned inside like a glove by the pressure of the incoming water.
The floor of the month and the skin of the throat are stretched by the pressure of the incoming water.

The animal has just opened its mouth. Water pressure turns the tongue inside out like a glove and stretches the floor of the mouth into an enormous pouch. A sphincter separates the buccal cavity from the pharynx.

The pressure of the water forces the tongue to turn inside out like a glove and the buccal floor to stretch out considerably to form the pouch that will contain it. Furthermore, this pouch expands between the skin and the body of the animal in the cavum ventrale by stretching the oral grooves pushing the hyoid apparatus and larynx upwards. The larynx will therefore be held firmly in place in the internal nostrils by the pressure of the water.

The articulation of the mandibles must be able to withstand the considerable tension resulting from the vast quantities of water taken in. It must also allow the mandibles to rotate when the mouth opens; it is thus substantially different from the articulation found in other mammals. It does not contain synovia, but instead an oil-filled cushion of fibrous tissue. This cushion is inserted between the GLENOID CAVITY and the articulatory extremity of the mandible.

The cushion can neither be stretched nor compressed; instead, it provides rotational movement of the mandible. It allows a rorqual to maintain a hydrodynamic shape when its mouth is closed and to increase the width of its mouth when it is opened.

Stranded Minke whale on an islet near Raguenau, Québec. The tongue and buccal floor have been distended by the gases released by the process of putrefaction. They have slipped out between the throat muscles and the buccal floor due to the presence of the prepharyngeal sphincter made by the palatoglossus. Notice the extent to which the pouch (made up of the tongue and buccal floor) has expanded.

Articulation of the mandible in the Minke whale. The oil-filled cushion (1) is clearly visible between the temporal process on the left (2), and the extremity of the mandible, to the right (3).

Photo mosaic showing the rotation of the mandibles towards the exterior at the opening of the mouth of a Minke whale.

The detailed functioning of the mouth is beyond the scope of this book; suffice to say that it shows an astonishing adaptation for capturing prey (Lambertsen, Ulrich, and Straley, 1995).

Minke whale feeding. The pink color of the belly is due to the presence of blood cooling itself in the skin.

We do not know exactly how the filtered food is transferred to the esophagus. We are still waiting for a volunteer to confirm the hypothesis that progressive waves generated by the contractions of the tongue muscles (a little like the peristalsis of the esophagus) are responsible for moving food along.

"Crawfoot," a Minke whale feeding.

The forestomach of a Blue whale can hold up to around a ton of krill. A Fin whale can have 800 kg (1,764 lbs) of food in its stomach. The stomach is many chambered, like the one in Odontocetes. Because cetaceans have to swallow enormous quantities of unchewed food, a many chambered stomach is advantageous. Swallowed prey items are stored and mixed without interfering with the digestive processes occurring in the next compartments.

Sandlance otolith in the forestomach of a Minke whale.

The pancreas, which produces the pancreatic enzymes necessary for digestion, is not diffused, but instead is remarkably compact, at least in the Beluga. The intestines, five to six times the body length for mysticetes, are not especially long, but do, nonetheless, reach 150 m (492 ft) in the blue whale. Maybe it is because mysticetes do not dive as deep and as long as Odontocetes.

Abdominal cavity of a Minke whale, Îles de la Madeleine, 2003.

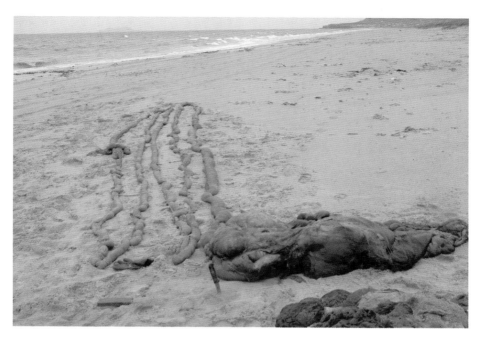

The intestines of a Minke whale are a lot shorter than those of an Odontocete, only five times the length of the body.

What do Mysticetes eat? They consume small prey items. The esophagus, even in the larger specimens, is no more than 10 to 15 centimetres (3.9 to 5.9 in.) in diameter. It would definitely have been difficult for such a whale to swallow Jonah!

Esophagus of a Minke whale, stretched to the limit, Îles de la Madeleine, 2003.

Some species, such as the Blue whale or the Right whale, feed almost exclusively on crustaceans that are superficially similar to shrimp but belong to the Euphausiacae or krill family—krill is a Norwegian word meaning "whale food." Since COPEPODS (*Canalus sp.*) are often found among the swarms of krill they are obviously swallowed as well, since sorting out the different species would be quite an achievement for these whales!

Blue whale feeding. Notice how the ventral pouch is distended.

Richard Sears

Retractor muscles of the oral (gular) grooves.

The other rorquals are much more eclectic. They will feed on krill, copepods (*Calanidae*), molluscs (sea butterflies, squid), small fish (capelin, smelt, sandlance, and Arctic cod), and the juveniles of bigger fish (herring and mackerel).

When foraging, rorquals do not display the same degree of synchronized behavior as seen in Odontocetes. Although these animals are not really gregarious, they have been observed gathering in areas where high concentrations of food are found.

Humpback whales, which sometimes form large groups, have developed a rather spectacular feeding method. Diving under a school of fish or krill, they slowly move up in a spiral while discharging a line of bubbles from their blowholes. The bubbles form a net that gathers in frightened fish or krill. (Underwater photographers know how hard it is to take pictures of fish because of the noise made when bubbles are released by the camera's trigger when shooting.) Then, the whales simply move up in the middle of the bubble-net, their mouth open, engulfing their food, and move on to repeat the sequence.

Sea butterfly. *Clione limacina*, also known as whale bait.

Many whales can cooperate during this type of feeding behavior, and the size of the bubble-net will vary with respect to the number of participants involved. Again, we see how nature has preceded man in its technology: we have learned how to catch a good number of fish species with bubble-nets created by submerged perforated pipes.

A Humpback surfacing, mouth wide open.

How much food do these giants consume? In 1976, Christina Lockyer published results from her work on the rorquals of the Antarctic. She concluded that they must consume about 4% of their total body weight daily, in the 120 dayfeeding season, in order to make up their annual metabolic needs. Migrating whales will spend the remaining time in warmer and less productive waters, and consequently eat much less. Over the year, then, Mysticetes eat close to 1.5 to 2% of their body weight each day.

How much time do they spend foraging? Is it all spent hunting or are other activities possible? After having estimated the sizes of fully stretched mouths of rorquals and assuming a krill density of 2 kg/m^3 (0.12 lbs./ft3) of water, Christina Lockyer calculated the number of gulps or mouthfuls needed to satisfy a whale's energy requirements. For instance, a 28 meter (92 ft) Blue whale weighing more than 120 tons needs 4 tons of food each day and will consequently need 79 gulps. We could try, for fun, to calculate the number of euphausids consumed, knowing that each weighs at the most 2 grams! A 22 meter (72 ft) Fin whale weighing 62 tons needs 73 gulps to get its 2 tons of krill it needs daily. On the other hand, a small 5 -meter (16 ft) Minke whale consuming its daily 711 kg (1,567 lbs.) of fish, needs 355 gulps. We can understand why these small cetaceans are so active at their feeding sites! But, most of all, we see here the advantage of having a great size. Obviously, as prey density increases, the number of gulps needed to feed decreases. It has been estimated (Moiseev, 1970) that krill concentrations may reach a density of 10 to 16 kg/m^3 (0.6 to 1.0 lbs./ft3). Once their forestomach is filled, a certain amount of time is needed for cetaceans to digest their food. Although they are less active during that time, digestion is relatively quick, thus maximizing feeding opportunities.

Cetaceans assimilate most of their food. A fin whale can assimilate up to 85% of its ingested food, which shows a highly efficient digestive system. It is not surprising then that a blue whale can put on as much as 49% of its body weight during the 120-150 days spent on its feeding grounds. This increase in weight is mostly in the form of blubber, but a gain in muscle mass is also observed. In addition, bones can store important quantities of oil, representing 10% the whale's total body weight.

Herd of Humpback whales feeding.

Minke whales violently expelling the water contained in their ventral pouches.

PINNIPEDS

Pinnipeds have also been modified by evolution to be able to be more efficient in capturing their prey in the water. Cheek teeth are homodonts and pointed, used only for grasping prey. Canines are usually well developed but incisors are fewer than in land carnivores and smaller. Milk dentition is made of incisors, canines, and premolars. Some species, like the Crab Eater seal (*Lobodon carcinophagus*) and, in a lesser way, the Leopard seal have additional points to their premolars and molars which are used to trap and strain krill.

The teeth of a crab eater seal.

The Walrus has a reduced dentition. Its cheek teeth have lost their points and have become crushing tools, used to crush the shells of the mollusks which are the main part of their diet. The canines are enormous, a lot bigger than what can be found in modern carnivores. They can be 1 m long and weight more than 5 kg. They have mainly social functions in male but can be used as weapons and to climb on ice floes by both sexes. It may be the kind of food it ate that allowed those canines to grow so much. As they did not impair the Walrus's ability to gather food and gave males lots of social advantages, natural selection favored the big tuskers.

The teeth of a walrus.

Otariids have the same cheek teeth as phocids, all alike and pointed, used only for grabbing prey.

139

The digestive system is similar to what is found in land carnivores. There are, nevertheless, differences, like in the length of the intestines. They are between 8 (California sea lion) and 25 times (Elephant seal) their body length. I was able to verify for a female Grey seal that they were 23 times her body length. This great length may be surprising as they do not eat very hard to digest preys. Maybe it is related to the fact that they do long and deep dives while foraging and that, during the apneas, the digestive system receives no blood. They have to maximize the efficiency of this system while on the surface or on land when it is irrigated normally.

Length of the intestines of a female gray seal.

Pinnipeds eat mainly fish or mollusks like squid. According to species, abundance or density of prey, they may forage both individually and cooperatively in groups of varying size. Some seal species, like the Leopard seal and the bigger sea lions can attack birds and other pinnipeds. When the prey is small enough it is swallowed whole, but if it is bigger it will be brought back to the surface to be eaten piecemeal.

The Walrus eats mainly shellfish. It detects them with the many vibrissae of its snout, and sucks them in directly or unearths them first with a powerful jet of water spat from its mouth. The tongue, pulled to the rear of the mouth by very powerful muscles, acts as a piston creating one atmosphere of negative pressure. This is quite enough to pull the bigger mollusks out of their shell. The smaller one are crushed whole. It was estimated (Fay, 1985) that an adult Walrus needs 6,000 shellfish a day and that it has to fish up to 17 hours in order to catch them. They can eat 6 of them in a minute! They eat seals, too, sucking in the skin, blubber, and viscera with their powerful tongue.

Some female pinnipeds fast during the whole period of lactation, others for only part of the same period. Males of some species also fast prior to reproduction to keep their rights on the part of the beach where they will attract their females.

MARINE ECOSYSTEMS AND FOOD RESOURCES

The ocean is an extremely complex ECOSYSTEM. In fact, it would be more appropriate to consider oceans as made up of many ecosystems. Although conditions encountered in the oceans are not as extreme as those experienced on the continents, the fact remains that there is a major difference between conditions encountered by organisms of the abyssal plain and those of the shallow areas of the Caribbean Sea. The immensity of the ocean often makes us forget its fragility and, like the rest of the planet, its vulnerability to the heedless actions of humans.

An ecosystem is defined as a complex arrangement of various plant and animal species, or BIOCENOSIS, that are grouped together by the unifying force of certain elements and physical conditions—air, water, temperature, climates, seasons, etc.—interacting within an environment called the BIOTOPE.

Within an ecosystem, organisms interact with one another in many different ways, interactions in which a transfer of both energy and matter takes place.

Perfection is not of this world, so imagining a perfectly self-sufficient structure working in a closed system is fruitless. For many years now, a host of scientists have tried to perfect perpetual motion but without success! The laws of thermodynamics explain in part the reason behind this situation.

Because it cannot operate entirely in a self-sustaining way, the ecological cycle needs a constant input of new energy and matter to keep it functional. Where does this energy come from? From the sun, which provides this energy by releasing light and heat on the entire planet.

Not all forms of life can use this type of energy directly. Only the plants that contain chlorophyll (a pigment responsible for their green coloration) can, through a process called PHOTOSYNTHESIS, produce their own food from the rays of the sun and inorganic matter (H_2O and CO_2). Since they can manufacture their own food, we call them AUTOTROPHS. These organisms then continuously introduce new organic matter in the ecosystems. Most other living things in the ecosystem will depend essentially on the autotrophs for their food. Such organisms are called HETEROTROPHS.

Whales are not vegetarians. But they do depend indirectly on the presence, in the oceans, of the hundreds of billions of plants so small that a microscope is needed to see them! These tons of PHYTOPLANKTON are used as food by the billions of tons of small animals, some microscopic, some bigger, which make up the ZOOPLANKTON.

Some whales feed directly on plankton. They are therefore close to the bottom of the food chain. They are relatively better off than the other cetaceans that find themselves higher up in the food chain.

The zooplankton usually serves as food for larvae or bigger animals, which in turn, will serve as food to even bigger animals, and so on; this sequence is part of the food chain. The higher an animal is in the food chain (i.e., higher TROPHIC LEVELS), the higher its risks are of accumulating toxic matter from the food it consumes, because this concentration increases with every step up the food chain.

For instance, let us imagine that a few grams of plankton are needed to make up an element of krill. If this plankton is slightly contaminated, each krill element will accumulate a negligible quantity of pollutants by feeding on this plankton. Still, the 4 tons of krill the blue whale takes up each day, even if low in contaminants, can have serious repercussions on its accumulation levels, and could even induce physiological disorders.

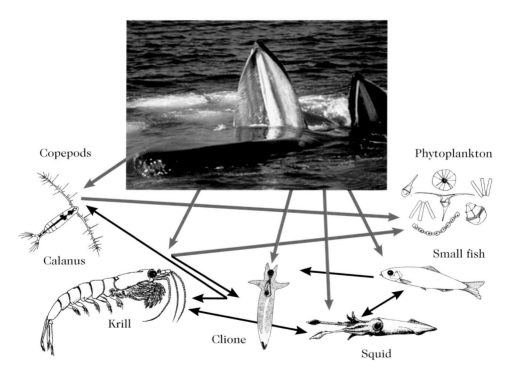

Food web involving a planktonivorous Mysticete. The arrows go from predator to prey.

If we now consider the case of a dolphin or a Killer whale, which feed on fish ... that have eaten other fish ... that fed on krill ... that had previously fed on contaminated plankton, we can clearly see that concentrations of pollutants increase with every step up the chain. Furthermore, if this chain unfortunately contains BENTHIC organisms that find their food in the sediments, then the problem is accentuated because burrowing organisms reintroduce contaminants into the chain that had been deposited at the bottom of the water. It is not without good reason that the Beluga population of the St. Lawrence River has had such a hard time recovering. These animals, particularly the females, feed on large quantities of *Nereis*, a predatory worm that lives in the sediments (Vladykov, 1946). Since Belugas are already in an area where sediments are highly contaminated, their feeding habits do not give them much chance to avoid contamination. These higher trophic-level animals, the top predators, may thus store important quantities of heavy metals and other contaminants in their tissues, and particularly in their fat. Every time they draw energy from their stored fat (e.g., from lack of food or during milk production in females), the contaminants recirculate through their body. Consequently, calves are being contaminated at the nipple by the fatty milk provided during their nursing periods. The immune system of the calves may then be weakened and their reproductive system affected.

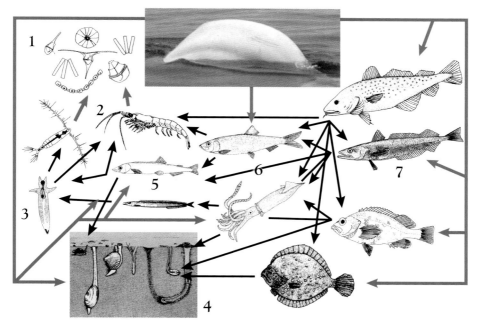

Food web involving a beluga whale. Arrows are from predator to prey. It is more complex than the web involving the Mysticetes since the beluga feeds at various trophic levels. 1) phytoplankton, 2) phytoplanktivores (phytoplankton eaters), 3) zooplanktivores (zooplankton eaters), 4) benthic organisms, 5) small fish, 6) squid and medium-sized fish, 7) large fish.

Sun, water, and carbon dioxide are not enough for green plants to complete their life cycle. Organic matter is needed to produce the different chemicals that make up their tissues and allow them to function. Excessive fertilization in agriculture, which today is destroying organic matter, is evidence of this need. Where do these natural fertilizers come from? They are produced from cadavers and other decaying organic matter. They end up on the soil or at the bottom of the oceans to serve as food for various decomposers, animals or plants. In turn, these organisms will also return part of the matter that they used to sustain themselves. Consequently, in the bottom of the oceans, we can generally find water rich in natural fertilizers. Water is continuously moved by different forces that create currents of varying strengths (the CORIOLIS force, the temperature of surface waters, tides, etc.). These currents collide with one another or with certain elements of the seabed's relief. In certain areas, these collisions create a great amount of turbulence, creating, in turn, an upward flow of deep water. The latter phenomenon is known as upwelling. This upward flow of cold, nutrient-rich water stimulates the growth of phytoplankton and, ultimately, the rest of the food web as well, especially when insolation, or solar irradiation, is optimal for plant growth.

The mouth of the Saguenay River is a home to this phenomenon, which explains the diversity and abundance of whales in the area. It is also unfortunately an area that has been highly polluted by the dumping of industrial waste over the last several decades.

When an animal produces the energy it needs, it will use organic elements in its food and metabolize them in presence of O_2. Marine mammals are all carnivores, with the exception of the Sirénians, the manatee and the dugong. They are continuously looking for prey. With the exception of the Right and the Blue whale which eat only crustaceans, and the Walrus, mainly a clam eater, marine mammals feed on fish and squid. Generally speaking, those fish need well oxygenated water to metabolize their food. One could think that seas and oceans contain O_2 in limitless quantities, so no fish should have problems with O_2. It might be true if fish were evenly distributed in the oceans, but it is not the case. They tend to gather in productive waters, very often near the coast. And it is there that the problems arise: Industrial countries practice intensive agriculture and livestock raising. It means the use of huge quantities of fertilizers and production of tons of manure, only part of which is used as fertilizer. The surplus of fertilizers and manure will eventually go in the rivers and from there to the sea where they will join up with organic matter coming from untreated city water and natural fertilizers. There, they will do their job: fertilize coastal waters mainly. This will promote a suproduction of organic matter (eutrophisation) which will eventually be broken down by specialists: the decomposers, which will liberate useful elements, but in so doing, they will use huge amounts of O_2. Eutrophized waters are usually poor in O_2, and this is not good for the survival of many species. We do not have to go very far to notice the effects of such output on a marine environment and its habitats.

It was recently noted that the waters of St. Lawrence maritime estuary contain a lot less O_2 than 75 years ago, even in the depths. The percentage of dissolved O_2 in waters 200 to 300 m deep went from 30% in 1930 to 17% today. This situation could have dramatic consequences, because all the fish that cannot tolerate low levels of O_2 could die if they go to feed in these impoverished waters. The cod, for example, like most fish, leave waters with a content of less than 20% of dissolved O_2 If fishes spend more than 4 days in such waters, 50% of them will die. One can imagine what will happen to all the deep living fishes which have to go there to feed. Apparently this zone extends from Les Escoumins to Matane. The mean temperature of these usually cold deep waters has risen by 1.7°C (which is a lot for such a big mass of water). This allows decomposers to be more active, and the more active they are, the more O_2 they use. As waters coming from upstream contain vast quantities of nitrates and phosphates (it passes 6 millions tons of these chemicals by Quebec City each year) it accelerates the eutrophisation process. The whole food web could be affected and that could have a very bad effect on marine mammals which are at the top of the food web and come to feed in this upwelling zone.

There is another place where one can easily see the effects of fertilizers on a marine environment: just look at the satellite picture showing the presence of algae in coastal waters of Canada (page 150). One can see the that the regions where the water is the coldest (upwelling zones are easily seen on the picture showing surface water temperatures. Page 150) are also the more productive. But it shows also that the sea south of Prince Edwards Island is very rich in chlorophyl, therefore in phytoplankton. It is not an upwelling zone, but a very good example of eutrophization of coastal waters by fertilizers.

Coastal habitats are essential for the survival of many marine species. It is of the upmost importance that we do our maximum to protect them by developing better methods to fertilize agricultural lands, controlling very severely the use of manure, and by treating the waters used by all the towns on the shores of the rivers and the sea.

Ursula Tscherter

MIGRATIONS

Migratory behavior in whales has evolved as a compromise between the need for more abundant food supplies and the need to reproduce. Migratory species usually move to the food-laden cold waters of northern and southern oceans in the spring. In early winter, they return to the tropical or temperate waters that contain far smaller food reserves, referred to as their wintering (and breeding) grounds.

These migratory patterns can involve considerable distances. California Grey whales travel from Alaska to the lagoons of Baja California, Pacific Humpback whales migrate from Alaska to the Hawaiian islands, while the North Atlantic Humpbacks travel from Greenland to the West Indies, on the Silver Bank near the Dominican Republics.

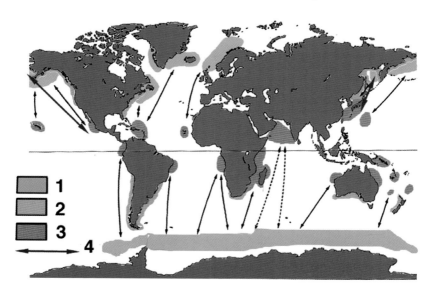

1. Feeding grounds
2. Breeding grounds *(Eubalaena glacialis, australis)*
3. Breeding grounds *(Megaptera novaengliae)*
4. Migration *(Eschrichtius Robustus)*

Migratory routes between feeding and breeding grounds of some cetacean species.

It may seem strange at first to learn that cetaceans leave their ocean "pastures," where food is plentiful, to travel to waters where food supplies are scantier and where they will live many months of almost total fasting. But the answer is not only simple; it also rests on a fundamental principle of nature. Simply stated, an animal has to spend energy (i.e., foraging) to get more of it (i.e., by feeding). If an animal invests more than what it gets in return, problems arise. Many of life's processes are function of costs versus benefits or of "return on investments." For instance, when finding food becomes too energy-consuming, animals will sometimes stop foraging and wait for more optimum conditions. Meanwhile, they can either enter into hibernation—an inactive condition usually characterized by a prolonged lowering of the metabolic rate—or they can migrate to their wintering grounds, where food is scarce but environmental conditions are more favorable. Once there, a whale can survive by living off its fat reserve. Moreover, the balmier conditions of their wintering grounds make bearing and nurturing young much easier. This is indeed what many cetaceans do. On the one hand, females give birth in warmer, albeit less productive waters of their winter breeding grounds and have to sustain long periods of fasting. The calf, on the other hand, feeds on its mother's milk and does not have to fight the rigors of the arctic waters. As a result, most of the food energy provided by the lactating female is invested in the calf's growth. When the time has come for the young whale to follow its migrating mother, it will be sufficiently large and strong enough to face the harsher conditions of the colder waters. The work of P. Corkeron (Australia) and M. Brown (South Africa) seems to confirm this hypothesis. While taking inventory of the wintering Humpback whales in the warmer tropical waters, they observed that the herds—one of the many names given to a group of whales—were made up of 75% males and 25% females. Since the sex ratio in this species is close to 50%, it is logical to assume that some females are missing. But where could they be? It seems likely that these missing females were probably sexually immature, very old, or not giving birth that year. Not having to undergo the trauma associated with calving, it is more advantageous for these females to stay in the colder waters, where the rigorous conditions are largely compensated by the abundance of food. The work of Corkeron and Brown may lead to a readjustment of the actual number of Humpback whales living along Australia's east coast, a number that had in fact already increased from 600 (in the 1960s) to an estimated 2,800 in 1995 (*Discover*, March 1996).

New tools used in cetacean research can facilitate the tracking of moving whales. An acoustic surveillance system (Sound Surveillance System: SOSUS) developed by the U.S. Navy is now used to track down and locate low-frequency sounds transmitted by some marine mammals. The easiest signals to detect appear to come from the vocalizations of Blue whales (*Balaenoptera musculus*). They produce sounds that can last 16 seconds in the 17-Hz range. By using special sound filters, we

Sequence of Pacific blue whale vocalizations recorded over a few days in September 1994 by SOSUS, an acoustic detection system.

- B94244
- B94245
- B94246
- B94247
- B94249
- B94251
- B94252
- B94254a
- B94256
- B94267

can isolate these sounds from the surrounding background noises and pinpoint the animal's position to within a few kilometres, even from a distance of many hundred kilometres. We can only rejoice to see military technology used in such peaceful ways! Consequently, we have accumulated, in a few months, more data on the position of Pacific Blue whales than in the many years of observations from the surface.

Where do whales go to feed? They go where food is both abundant and easy to get. Three types of situations fulfil these requirements.

First, they feed in areas where two water masses of different temperature meet.

Second, areas where eddies are formed by colliding currents, or by a current deviated by a land mass or by sediments, are potential feeding grounds.

Plankton accumulates in high densities in these areas. Animals that consume plankton or depend on them indirectly for food will also congregate in these small areas.

The third kind of feeding zone is found in areas where the deep, generally colder and nutrient-rich water rises toward the surface. This vertical, upward movement, appropriately called upwelling, can be dynamic (i.e., a product of the interaction of different currents and atmospheric movements such as winds), or topographic (i.e., when deep currents encounter obstacles such as oceanic ridges or submarine mountains). These waters are rich in minerals (i.e., nitrates and phosphates), which are returned to the environment by the decomposers (bacteria) that feed on the organic debris that drizzles down from the surface and settles on the ocean floor. These phosphates and nitrates act as fertilizers that stimulate massive production of *phytoplankton*, the vegetal organisms of the plankton, which fuels all trophic levels of the oceans (Gaskin, 1982).

As noted above, one such area exists in the St. Lawrence River, near the mouth of the Saguenay fjord (Québec, Canada).

149

Cold and productive waters (5°C).   Warmer and less-productive waters.

Satellite image of the estuary and the gulf of St. Lawrence showing temperatures of the surface water. Zones of upwelling are indicated in blue.

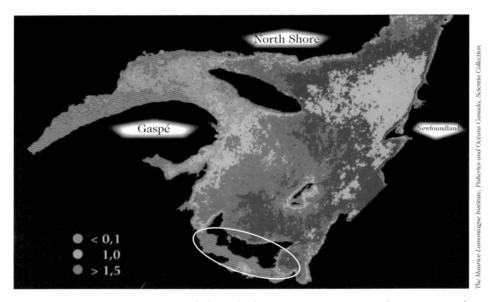

Collection Mean concentrations of phytoplanktonic pigments (microalgae) in mg/m³. In the oval, we can see a very productive area, in spite of high water temperature. Could it be the effects of agricultural pollution?

A synthesis of satellite images showing the presence of algae in the surface waters. A strong correlation exists between the zones rich in algae and those of upwelling.

Recording of a krill swarm in the deep trench in the St. Lawrence Estuary off Les Escoumins (North-Shore, Québec). It is approximately 15 km wide and 100 to 200 m thick. Recordings were made using a high-frequency acoustic echosounder (120 kHz) on August 1st 1994 by Fisheries and Oceans Canada.
Yvan Simard, Maurice Lamontagne Institute

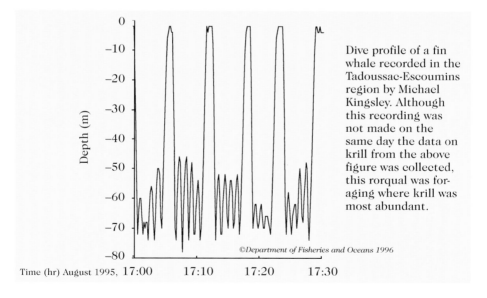

Dive profile of a fin whale recorded in the Tadoussac-Escoumins region by Michael Kingsley. Although this recording was not made on the same day the data on krill from the above figure was collected, this rorqual was foraging where krill was most abundant.

The cold waters of the Laurentian Channel, where depths exceed three hundred meters (985 ft), collide with the shoals found near the mouth of the Saguenay River. Little is known about the exact causes behind this movement of water because many factors are involved: tidal currents, the influence of the cold waters of winter that possibly downwells come spring, the possible role of the cold Labrador current. But, however obscure these causes are, the results are nonetheless very clear: an upwelling of fertilizing waters that fuels the proliferation of microscopic algae of the phytoplankton, which in turn serves as food for the zooplankton such as copepods and euphausids. These animals are not necessarily from that region, however. Currents might easily carry the zooplankton there and the conditions so rich in organic life that characterize this part of the St. Lawrence could contribute to their staying. As result, the zooplankton could thrive and serve as food for the many fish and ultimately the whales. Therefore, it is easy to understand why there are so many whales there and why this region is known as one of the best whale-watching sites in the world.

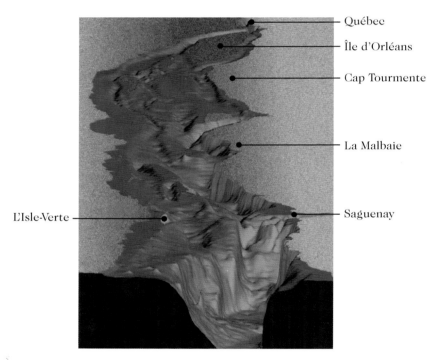

Perspective image of the Laurentian channel. This 3-D image is a synthesis of the bathymetric data of the St. Lawrence's upper estuary. *The Maurice-Lamontagne Institute, Fisheries and Oceans Canada, Scienta Collection.*

Pinnipeds

Some species of pinnipeds can also migrate long distances. But, instead of looking for warmer waters to have their young, they will be looking for places were they can haul-out to molt and lay down their pups. All phocids have an annual molt and during this time they have to be on dry ground, because there is much blood circulating in their skin (to grow the new fur) and it would be too energy costly for them to spend long time in the water. All pinnipeds lay their young on land and must find suitable places to do so, and these places are not necessarily the best ones to find food. It does not imply migrating every time. We have to distinguish between local movements, haul-out sites and foraging sites (even if it takes several days to do the round trip), and migrations which are long trips in time and space from one place to another.

Among phocids, the elephant seals can migrate more than 4,000 km between foraging and haul-out sites. They even do that twice a year: once to breed and once to molt, totaling nearly 20,000 km. Hooded seal can do 1,000 km, and Greenland seals can swim more than 4,000 km going from lat 70°N where they feed to lat 40°N where they have their young on the ice. The Grey seal from the St. Lawrence leaves the estuary and gulf in November to go, primarily, to the ice of the strait of Northumberland and on Sable Island, off of Nova Scotia, where they have their young. They come back to the estuary in may.

Planktonic elements, especially krill, are rich in proteins. Human beings have just recently discovered the enormous potential that krill offers and certain countries have begun to exploit this resource. Let us hope they will use it rationally, and that the competition for krill between humans and whales will not be added to all the other problems that already threaten the survival of certain species of cetaceans.

We may allow ourselves a degree of pessimism concerning this hope. In fact, as long as the sole concern of those who exploit flora and fauna is short-term maximum profits, as long as the government's only concern is to buy social peace (and get re-elected), we will relive the drama of seeing the near-extinction of these exploited species. Moreover, the necessary protective measures will, in all likelihood, be unpopular ones, since they will assuredly bring about a reduction in profits and will remain unpopular even if, in the long-run, they turn out to be more profitable to both exploiter and exploited. We will thus continue to see the forest disappear, to see fish such as the Atlantic cod brought close to extinction, and others, *ad nauseam*. Woe betide cetaceans should a crafty individual find a really profitable way to extract proteins out of krill, woe betide any creature that depends on krill. It is hard to believe that krill-exploiting factory ships will content themselves with functioning the way skimming whales do!

Shores and Tidal Flats

Some habitats are fragile and essentials, and we must protect them at all costs if we want to maintain the salubrity of coastal waters. These are the shores and tidal flats, the regions covered and uncovered by the tide. The shores of the sea have always exerted a great interest on humans who have built their harbours and have used them not too wisely for their summer resorts. The many towns and villages which were built on the shores of the St. Lawrence have dumped in it their sewage without any treatment. When the population of these towns and villages was not too big, and that the number of the industries they contained were not very important and worked on a small scale, the capacities of the environment to absorb and recycle the waste, mainly organic at the time, were enough to maintain it in good health. But the towns and villages became bigger, farming developed new methods, becoming almost industrial, and industries diversified. Their needs for water grew, and water containing more and more toxic waste was dumped, still without treatment.

The Saint Lawrence Maritime Estuary and the Gulf receive all the pollutants produced from the Great Lakes. Progress has been made and people have begun to give attention to a very precarious situation. The big cities, at least in Canada treat their waste water and industrial wastes better and better, The smaller villages do not treat their sewage because they do not have the financial means to do so, and a good number of citizens who have vested interests continue to pollute without remorse.

The shores, especially in semi-swampy areas covered and uncovered by the tide, play an essential part in purifying and recycling of waste contained in the water. A great number of fish come there to reproduce and wild fowl find food and protection there, as do marine mammals. Every time that man, by his actions, affects those habitats, to augment tillable land, to build cottages, or for harbour installations, he endangers the quality of the coastal waters, the most productive, the most necessary element for the stability of marine ecosystems. Every time a "garbage can" tanker is left navigating it constitutes an awful risk, particularly in the areas where the tide recedes on very large tidal flats. It is virtually impossible to clean such places when there is an oil spill and the oil is deposited there by the lowering tide. We are just beginning to measure long term effects of oil in the environment by studying sites where there have been big oil spills in the past, in America and Europe. We have to be very firm toward our governments so that they will pass and enforce very strict measures concerning the way we utilize the resources of the sea. We have the duty also to clean up our own back yard, by applying to our houses and cottages, all the rules made to protect the environment, even if our immediate interests seem to dictate that we dispense with the rules.

OSMOREGULATION

Cetaceans, with the exception of river dolphins, are all marine animals. When feeding, they can voluntarily or accidentally ingest considerable amounts of seawater, which contains a significant concentration of salt (an average of 35 g/l or 3.5% or 35%).

They thus find themselves in a situation similar to that of a castaway at sea trying to combat thirst by drinking seawater, a potentially fatal action. Most people are unaware of the explanation of this phenomenon. There are two reasons:

• First, by drinking seawater, the castaway's body fluids are greatly altered. The increased amount of mineral salts absorbed by the body substantially modifies the composition of his body fluids and affects cellular function. He must then eliminate excess salt by urinating. Unfortunately, in humans, water makes up most of the urine. Furthermore, human kidneys are incapable of producing urine that has a higher salt content than seawater. Consequently, to eliminate 1 litre of seawater, our castaway must produce 1.35 litre of urine. Naturally, dehydration becomes unavoidable.

Bladder of a Blue whale, Anticosti, 1987. This photo illustrates the difficulties in performing necropsies on such giant animals.

Reniculate kidneys of a Beluga. The reniculi are clearly seen on the right kidney. The fibrous coat that normally envelops them has been removed.

• Second, seawater has large amounts of magnesium and sulphate. These two substances act as laxatives for humans and cause diarrhea, which further accelerates the dehydration process.

Are cetaceans confronted with similar problems?

As we have mentioned, they feed on organisms with high salt content, and undoubtedly must ingest some seawater.

Studies have shown that the cetacean kidney produces urine that is more concentrated than seawater. A whale only needs to produce 0.650 litre of urine to eliminate the equivalent of 1 litre of seawater. As a result, it not only avoids dehydration but purifies and retains one-third of the seawater ingested (Schmidt-Nielsen, 1975).

No one knows exactly how cetacean kidneys work so efficiently. They are relatively and absolutely very large—the weight of cetacean kidneys is equivalent to twice that of a terrestrial mammal of equivalent weight (Slijper, 1976) and can reach 200 kg (441 lbs.) in a 60-ton Fin whale.

They are divided into small lobules called reniculi that presumably function as independent miniature kidneys, a characteristic shared with the other marine mammals and some ungulates. The number of reniculi is substantial, giving the cetacean kidney the appearance of a gigantic bunch of grapes. There are up to 3,000 reniculi in the Blue whale.

Reniculate kidneys of a Minke whale.

Diagram of One Part of the Renal Filtering System

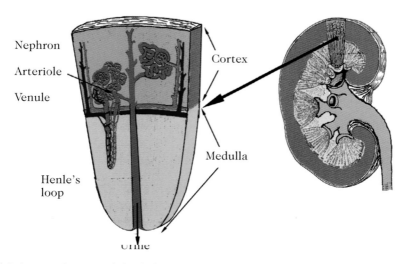

Henle's loop is the part of the kidney in which urine is concentrated.

157

The partition in reniculi of the kidneys of this juvenile Harbour seal is conspicuous (the scalpel points to the suprarenal gland)

This division into reniculi shortens the structure generally associated with the ability of kidneys to concentrate urine, a structure known as Henle's loop.

Desert mammals that produce highly concentrated urine usually have very long Henle's loops. Cetaceans, however, have very short ones since their kidneys are divided into reniculi that considerably diminishes the thickness of the part that contains them. This characteristic sets them in a rather distinctive group. We will have to wait until more information is available concerning this subject in order to solve what appears to be yet another paradox (Beuchat, 1996). This could be a pseudoparadox, if marine mammals have a way to make up with their need for water in other ways than drinking seawater.

Marine mammals as we have said, drink sea water, voluntarily (mariposia) or not (some is swallowed accidentally with the preys), but they also have other means to pick up the water they need. In addition of producing a more concentrated urine (Elephant seal pups can reduce their urine output by 84% after 10 weeks of fasting [Adam and Costa, 1993]) they can recover up to 1.07 g of water when they metabolize 1 gram of fat and 0.4 g of water for each gram of proteins. The water content of their prey varies from 60 to 80%. All this water can meet a good part of their needs. The fatter the prey the less they need to drink! When they fast, by using their fat reserves they not only meet their energy needs, they quench their thirst! This process is so efficient that a sea lion was able to spend 45 days without water without any adverse effect.

Seals limit also their water loss by a counter current exchanger system. Their nostrils contain many very complicated folds of bone on which there is skin which has a very big surface and contains many blood vessels. The moisture from expired air condenses on this skin and is recovered.

Ethmoturbinates of a Greenland seal.

Lactating female whales must also guard against water loss. To limit loss, they produce milk that, compared to the milk of terrestrial mammals, is much more concentrated. Changes in the proportions of the milk constituents (fat, proteins, water, etc.) may also be observed during lactation in marine mammals, such as the Weddell seal. The water content in this seal's milk is, on average, 43.6%. It can be as low as 27.2% to allow the female to give sufficient food to the pup without dehydrating. In such instances, this milk contains 57.9% fat and 19.5% protein.

ORES – Ursula Tscherter

REPRODUCTION

Cetaceans are PLACENTAL mammals with internal fertilization. In this respect they are not any different from terrestrial mammals. The life of a 130-ton Blue whale starts as discretely as that of a shrew, which weighs just a few grams, by the union of a spermatozoid and an ovum, which are incidentally about the same size for both species. However, reproduction and all associated activities are admittedly a little more spectacular in cetaceans!

We have already seen how difficult distinguishing a female from a male is, because the latter lacks a scrotum and its penis is normally completely retracted, while the teats in nonlactating females are hidden deep inside individual folds of skin. The position of the anus in relation to the genital slit is frequently the only criterion that enables us to discriminate between the sexes. In females, the anus and the genital (or vaginal) slit are included in the genitoanal slit. In males, however, the genital or penile slit is located halfway between the umbilicus and the anus. It is approximately in the same position in pinnipeds A visible and easily identifiable sexual dimorphism is present in only a few species. In the Killer whales, for instance, the dorsal fin of males is larger and more triangular, whereas in females, its posterior margin is crescent-shaped.

Male Killer whale in the foreground with a female in the background. The difference in the shape of the dorsal fins is clearly visible.

Michael W. Newcomer

The male Narwhal is also easily recognizable, by its long ivory tusk (the left incisor), which earned it the nickname "sea unicorn." Visible teeth in the males of adult beaked whales allow us to distinguish the sexes.

Conspicuous tooth of the male Narwhal.

MALE GENITALIA

The penis is composed of erectile tissues enclosed with a fibrous sheath. It is kept inside the body by a bundle of loose retractor muscles that are attached in part to the rectum. There is little lengthening of the penis during erection but, like the ruminants, it does straighten. A certain degree of rigidity in the penis is ensured during coitus through the combined action of the sheath's elasticity and blood-filled erectile tissues. When a whale beaches, the retractor muscles relax, causing the penis to protrude from the carcass.

Ombilic (O), genital slit (OG), and anus (A), Harbour seal, juvenile.

Genitoanal area, Grey seal female: vagina (V), anus (A).

Female Harbour porpoise: We cannot distinguish the anus from the genital slit.

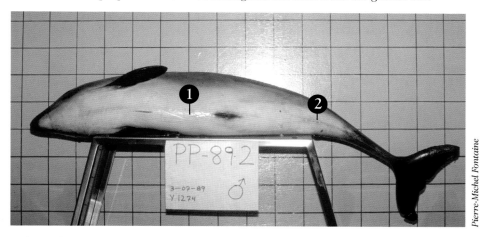

Male Harbour porpoise : the genital slit is between the umbilicus (1) and the anus (2).

The erectile tissues consist of a cavernous body (*corpus cavernosum*), which splits in two at the base of the penis and attach themselves to the vestigial bones of the pelvic girdle, and a rudimentary *corpus spongiosum*, which surrounds the urethra.

Proportionately speaking, the penis of a whale is no longer than that of ruminants. Still, the Blue whale's penis can reach lengths of 3 m (9.84 ft) and have a diameter of 30 cm (0.98 ft)! It lacks a baculum, or os penis, found in pinnipeds, dogs, or bears, for example.

Penis of a Sperm whale. The relaxation of the retractor muscles has allowed it to come out of its slit.

Baculums of a Walrus (top) and a Grey seal.

A preputial membrane covers the tip of the penis. It is said that this membrane was the only skin of the great Mysticetes that could be tanned; whalers prized it for its use in making supple clothing. The Peabody Museum in Salem, Massachusetts, had on display a megaphone made from the preputial skin of a cetacean.

164

Cross section of a Blue whale's penis. 1. Urethra and corpus spongiosum; 2. Cavernous body; 3. Fibrous sheath.

Megaphone made from the preputial skin of a cetacean. Peabody Museum, Salem.

    The testicles rest on the dorsal wall and can be very large. In fact, they can reach 4 to 5% of a Harbour porpoise's total weight—more than one kilogram (2.2 lbs.)—a mammalian record. This huge size is probably due to a reproductive strategy based on the quantity of spermatozoids produced (Fontaine & Barrette, 1997). The testicles of the Blue whale can attain a length of 80 cm (2.6 ft), each weighing 45 kg (99 lbs.; Slijper, 1962). But next in line for the record held by the Harbour porpoise is the Black Right whale. Its total testicular weight is approximately one ton and each testicle measures 2 m (6.6 ft; Cummings, 1985)! In seals they are just under the blubber and lie lateral to the penis, and otariids have them in a scrotum.

Ann Pabst and Bill McLellan from the University of North Carolina have just recently discovered that cetacean testicles are surrounded by a complex venal and arterial network that functions as a countercurrent heat exchanger. This network maintains favorable conditions for spermatogenesis. In fact, cetacean testicles—being internal, close to the swimming muscles and insulated from the environment by a layer of blubber—are subjected to temperatures that would be high enough to interfere with the normal production of viable spermatozoids, especially in an active animal. To ensure proper cooling, a network of veins from the dorsal fin brings cooler blood in close contact to the testicular arterial blood. The resulting countercurrent heat exchange cools the testicles and normal spermatogenesis can take place. In pinnipeds it is the blood from the feet that cools them.

Harbour seal, male genitalia: Penis (P), Bladder (B), and Testicles (T).

Genitalia of a male Harbour porpoise. 1. Innominate bone; 2. Testicles; 3. Penis; 4. Retractor muscle.

White-sided dolphin: testicle with its cooling system.

The prostate gland, which is the only accessory gland of the male reproductive system in cetaceans, is huge in some species, like the Harbour porpoise. This highly muscular gland is located between the two branches of the cavernous body.

Since the prostate gland secretes most of the liquids found in the ejaculate and considering the high number of coitus during the breeding season, its relatively and absolutely enormous size is not at all unexpected.

Prostate gland of a Harbour porpoise. 1. Testicle; 2. Prostate; 3. Penis.

As mentioned earlier, cetaceans (or, at least captive dolphins) use their penis as a tactile organ in exploring their environment. Males frequently masturbate by rubbing their genitals against floating objects or the walls of the aquarium. Spermatozoids are a little shorter than human ones.

167

FEMALE GENITALIA

The ovaries are dorsally placed in the abdominal cavity close to the bicornuate uterus, a position somewhat like that of the testicles in males. The ovaries of Odontocetes are similar in appearance to those of other mammals. The surface of the ovaries in Mysticetes is bloated by FOLLICLES at various stages of maturation, sometimes showing a CORPUS LUTEUM (yellow body) with CORPUS ALBICANTIA (white bodies), giving them the appearance of a bunch of grapes.

The corpus albicans of cetaceans does not disappear following ovulation, but remains in the ovary probably throughout the animal's life. It is a scar left in the ovary by a corpus luteum. Since the corpus luteum develops in the ovaries only following ovulation, one simply has to count the corpus albicans to know the number of ovulations the animal went through in its life. Still, this method is complicated because many follicles can show certain signs of maturation during the same cycle and consequently leave many scars in the ovary. Because of this, accurate interpretation of the number of cycles is difficult.

Diagram of a Mysticete Ovary

Adapted from Slijper.

Beluga ovary. Three corpus albicans are visible at the surface. Two on the other side have been dissected.

Genitoanal slit of a Beluga. The tip of the measuring tape is near the clitoris and the vaginal orifice.

In Odontocetes, gestation takes place, 80% of the time, in the left side of the uterus, which leads us to believe that the left ovary is the more functional of the two. In Mysticetes, both ovaries appear to be equally functional.

The ovary of a Blue whale can weigh, depending on the size of the corpus luteum that it contains, between 8 and 29 kg (17.6 and 63.9 lbs.).

The two horns of the uterus unite to form a uterine cavity held shut by a powerful CERVIX.

The vagina is composed of the clitoris, a type of hymen in virgins, and folds that project outward like a series of funnels. It is believed that these folds either prevent water from entering the vagina or retain sperm after copulation. The orifice of the vagina lies deep within the genitoanal slit. The teats, also located within slits in nonlactating females, are on either side of the genitoanal slit. In some Odontocetes species, the female genital structure includes a vaginal cap, a highly unusual structure composed of secretions from the vaginal walls. Containing hard-calcified substances, its role is problematic. It is never found among female Mysticetes.

The uterus is positioned, somewhat as males' testicles are, between two muscle masses that produce great quantities of heat when the animal is active. And, as is the case for spermatogenesis, important reproductive processes such as

Female genitalia of a Beluga: 1. Vagina; 2. Uterus. The ovaries are shown in the white circles.

Female genitalia of the Minke whale. Even in this immature specimen, the extensive network of blood vessels that function as a counter-current heat exchanger during gestation is seen.

normal fetal development in gestating females would be jeopardized were it not for the activity of the countercurrent heat exchanger in the female genitalia. In October 1996, while dissecting an immature female carcass on île Verte, we noticed the presence of numerous veins on the ligament that holds the uterus against the abdominal wall. These veins contain blood from the dorsal fin and from the skin that is cooled by the surrounding water. This blood flows in opposite direction to the arterial blood originating from the dorsal aorta. By the time the latter reaches the uterus, it has been cooled by the venal blood, by means of the same countercurrent heat exchanger principle mentioned in the section on conductivity. This way, a more compatible temperature for the normal development of the fetus is maintained.

The genital organs of pinnipeds look a lot like those in land carnivores. Ovaries are situated near the kidneys and the uterus is bicornuate. It is cooled by the blood returning from the feet.

Gray seal female genitalia: ovaries (O), bladder (V), uterus (U).

MATING

Cetaceans go through long and highly visible courtship displays prior to mating. Furthermore, contrary to most mammals, in which olfactory cues are used to assess the female's receptivity and readiness to reproduce, cetaceans, particularly Odontocetes whose olfactory ability is totally lacking, must rely on their other senses, such as vision, touch, audition, and perhaps even taste through their Jacobson organ. The mating rituals of cetaceans are long and sometimes spectacular. A great deal of body contact is involved. Mutual rubbing, slapping, nuzzling and nibbling, sometimes lasting many days, are observed. Some species such as the Humpback whale can propel itself out of the water (breach), hit the water with its tail (tail slap) and its fins (pec slap), as a demonstration of strength and power. Competition among males can turn into intense and brutal confrontations. Besides the usual head-buttings and tail-slappings, the male Black Right whales will use the barnacle-encrusted callosities on their head to inflict severe wounds on their rivals. In Odontocetes, teeth are also used as weapons. Only male beaked whales have visible teeth, and the scars that they bear are indicative of their purpose. The broken mandible of Sperm whales is not always the work of clumsy whalers but rather evidence of the violent battles that can occur during the rutting season.

The mating period is inversely proportional to the exuberance and length of the courtship ritual: it lasts about five to twenty seconds. The partners most often swim alongside one another, belly to belly, and the penis, much like a homing device, finds the vagina after very little exploration, and briefly penetrates it. This activity is not at all easy, considering that the penis and the vagina are more than 10 m (32.8 ft) from the head, that the whales are in water and that they have no arms to stabilize them, although the caudal fin is often used by the male to this end. As seen in their ruminant cousins, coitus is brief but can be repeated many times. Cetaceans, as mentioned earlier, have a large prostate gland and can store huge amounts of sperm in the EPIDIDYMIS and VAS DEFERENS.

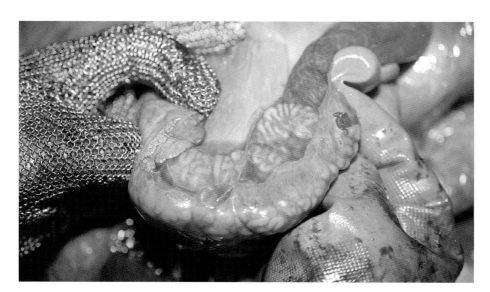

Epididymis of a Harbour porpoise.

It is not surprising that the first humans to witness the mating of Sperm whales thought that their head was filled with sperm. The size of their head and organs plus the frequency of mating clearly impressed those observers! It is for those reasons that these whales were called Sperm whales and the oil found in their heads, spermaceti (*sperma*: sperm, *ketos*: whale).

GESTATION

Despite whales' large size, their gestation period seems abnormally short. Most carry their young for 10 to 12 months depending on the species, with the exception of some Odontocetes, such as Pilot and Sperm whales, whose pregnancies can last from 15 to 16 months. The gestation period in mammals is usually linked to three factors: the size of the mother, the degree of development in the newborn at birth, and the number of young. A mouse, after 21 days of gestation, gives birth to 12 to 14 blind and hairless young, which then require further time to develop. Likewise, even though a human being takes 9 months to gestate, additional and substantial time is needed after birth to complete development, probably that of the human brain, unique in the animal kingdom. A horse's gestation period is 11 months, a camel's is longer still, and an elephant's is 22 months. Mysticetes' relatively short gestation period is probably linked to the migrations that they must undertake between the food-laden but colder waters and their winter calving or breeding grounds. They are extremely efficient at consuming high-energy food and can probably invest considerable energy in their fetus without interference to their ability to store fat, the crucial energy source needed to accomplish their migration and provide for their calf at birth. A rapid fetal growth is also, in all likelihood, more economical in terms of energy than a slower one (i.e., longer gestation) spread over two migration cycles and including a long fasting period. It is worth noting that the Sperm whale, which has the longest gestation period, does not migrate considerable distances and feeds on squid, a relatively lower energy food when compared with krill.

The cetacean placenta is cotyledonary, like that of hoofed ruminants. Minke whale, October 1997.

Fetal growth is extremely rapid, especially in the last two months of gestation, during which, for instance, the fetus of a Blue whale gains approximately 2 tons! The energy needed to produce—in 11 months—a calf that weighs 2.5 tons out of a fertilized ovum that weighs 0.005 mg (1,000 mg = 1 kg, 1,000 kg = 1 ton) is positively mind-boggling.

Fetus of a Minke whale. The mother beached on October, 1997, after having been hit by a ship.

Small cetaceans, such as the Harbour porpoises, which weigh on average between 50 and 60 kg (110 and 132 lbs.), seem to have a relatively long gestation period lasting 11 to 12 months. Moreover, at birth, calves appear proportionately larger than those of bigger species. In fact, by birth, they already measure more than one-fourth the size of their mother and weigh one-tenth as much. The explanation for this longer gestation period lies in the fact that a small animal has a larger, less favorable surface to volume ratio. As we noted in the discussion on conductivity, it has proportionately more skin than a larger animal. The Harbour porpoise, already the smallest of all cetaceans is then at an even greater disadvantage. To compensate, its calf must be allowed to develop as much as possible before birth. This comparison in relative birth sizes helps us to appreciate the advantages conferred by larger sizes. The calf of the Blue whale measures 7 m (23 ft) at birth (less than one-fourth its mother's length) and weighs 2.5 tons (over 60 times less). It is thus proportionately and absolutely less strenuous for a Blue whale to complete the fetal development of a 2.5 ton calf over the same period of time, 11 months, that it takes a porpoise to deliver a 5 kg calf (11 lbs.).

Stranded pregnant Beluga: the calf is still in the uterus. There is no visible cause to explain its death.

Stranded pregnant Beluga: the calf size is impressive (1.54m). The female measured only 3.5m. Was the calf proportionally too big?

    Calving is usually done tail first and the umbilical cord breaks, on its own, near the umbilicus, where there is less resistance. This tail-first birth is quite logical even if it does not appear to be at first, since it occurs in water, where the calf cannot breathe. The stimulus that initiates the need

Lengths and Weights of Some Cetaceans at Birth

Black right whale:	4.5 to 6 m (14.8 to 19.7 ft)	1 ton.	(Adult: 30 to 80 tons)
Blue whale:	7 m (23 ft)	2.5 tons.	(Adult: 100 to 130 tons)
Fin whale:	6 m (19.7 ft)	1.8 ton.	(Adult: 30 to 90 tons)
Humpback whale:	4 to 5 m (13.2 to 16.4 ft)	1 to 1.5 ton.	(Adult: 25 to 40 tons)
Minke whale:	2.4 to 2.8 m (7.8 to 9.2 ft)	0,350 ton.	(Adult: 5 to 12 tons)
Sperm whale:	3.5 to 4.5 m (11.5 to 14.8 ft)	1 ton.	(Adult: 20 to 55 tons)
Beluga:	1.5 to 1.6 m (4.9 to 5.3 ft)	80 kg.	(Adult: 0.4 to 1.5 tons)

to breathe can come from a change in temperature, which could have disastrous consequences if the head was the first part of the whale's body introduced to its new environment. Furthermore, and contrary to cows and horses for example, the heaviest part of the newborn whale is not the hind but the fore section. It is quite normal, then, (and coherent with the fetal position of terrestrial mammals) that the heaviest part of the body be as close as possible to the center of gravity of the mother. When the calf is born, its mother, usually assisted by other females, pushes it toward the surface for its first breath. The placenta is eliminated from the mother soon after and the calf rapidly starts to suckle.

PINNIPEDS

Pinnipeds use various strategies when they mate. The gestation period seems to average 11 months, with the Walrus as an exception at 15 months. In fact, the real gestation period averages 8 months (10 months for the walrus), because, after fecundation, the embryo will develop only to the blastocyst phase then enter a *diapause* of 3 to 4 months. Most of the seals species copulate in the water, except the Grey seal, the Crabeater seal, and the Elephant seals, but all the births occur on land or ice. All otariids copulate on land. *Polygamy* is the rule among pinnipeds. Among the species which copulate on land, dominant males will defend territories or groups of females they have gathered or which have gone to them. There is a big advantage for female otariids, for example, to reassemble themselves in the territory of a dominant male or to be a member of a group defended by a dominant male. In this way, they are protected from harassment from non-dominant males less fit to reproduce, which could have an impact on the care females can provide to their pups, Male otariids, which have maintained a better mobility on land than phocids, generally defend territories that have qualities looked for by females (like shady places for staying cooler). Seals, like the Grey seal, for example, would rather defend female groups because it usually implies less moving.

It is easier for pinnipeds to know the state of receptivity of the females, since they have their young on land and they become receptive after parturition or weaning of the pups.

In some species, as we said, males defend territories or groups of females and it can be the source of rather violent combat. Elephant seals in phocids, and otariids, generally, gather in very big groups. One male can have access to more than 100 females. The nearby males have to content themselves with catching females near the boundary of the territory or at the edge of the group. Reproductive success of non-dominant males is very low in these cases. For the seals that mate in the water, fights are usually rarer and less violent. The males will use vocal and visual displays and stay near the places where the females are nursing to have their chance to pass on their genes.

LACTATION

Cetacean milk, especially in Mysticetes, is extremely rich. Its caloric value is 4,137 kcal/kg (1,880 kcal/lbs.). In comparison, cow milk has only 745 kcal/kg (338.6 kcal/lbs.). Blue whale's milk contains 36% fat, 13% proteins, and 14% minerals, the remainder being water. Cow milk is 90% water, 3 to 4% fat, 4 to 5% solids (minerals, proteins and sugars). Cetacean milk has a creamy consistency, a little like milk concentrate. I have tasted milk from a Beluga and found it thick with a slight salty aftertaste, but not at all unpleasant.

The female Odontocetes do not fast while nursing. Their milk contains a lot less fat than that of Mysticetes (less than 30% usually). They compensate by the length of nursing, which is usually more than one year, and the constant availability of the mother which is less stressful. There is a notable exception among Odontocetes: the Harbour porpoise (*Phocoena phocoena*) nurses its calf only 8 month but with a peculiarly rich milk (46% of fat). It is certainly to compensate for the very small size of the calf.

Nursing is initially frequent and of short duration since the calf can remain under underwater a short time only. With time, nursing intervals become gradually longer. To nurse, the mother lies on its side while the calf, from the tip or side of its mouth, grabs the protruding nipple. It does not actively suckle. In fact, once the calf has the nipple in its mouth, special mammary muscles that surround a type of reservoir contract and vigorously force milk into the calf's mouth.

Such a procedure is an adaptive advantage. In a mobile environment such as the ocean, briefer nursing periods lessen the risks of interruption while feeding and ensure an appropriate milk ration.

Open mammary glands of a Minke whale. The finger to the left is in that part of the reservoir leading to the teat. The right hand points to the open reservoir.

Section through the mammary gland of a Harbour porpoise. The yellowish milk is seen in its reservoir.

Milk flowing from the teats of a female Northern Bottlenose whale stranded near Montmagny, November 1995.

Among pinnipeds we find the females spending the less time nursing their young. The Hooded seal is the real champion since the female nurses its pup only 4 days! The Greenland seal nurses 12 days, the Grey seal 16, and the Harbour seal 24 days. It is the Weddell seal that nurses the longest time, 53 days. Otariids, by contrast, nurse their pups between 365 and 540 days. Most of the pinnipeds feed themselves during lactation, or at least part of it. Those that fast during lactation are mainly the big species. Their superior mass allows them to accumulate the reserves necessary to produce the rich milk during the relatively short time of nursing. The fat content of the pinnipeds is in an inverse ratio to the length of nursing. The fat content of the Hooded seal milk is 60%, like that of the Grey seal. Weddell seal milk has only 48% of fat.

Otariids milk does not follow this rule, which is surprising at first glance. In spite of a nursing period of one year, the California sea lion, and the Falklands sea lion (*Arctocephalus australis*) milk have 44 and 55% fat respectively. This is more easily understood if we consider that the females spend a long time at sea feeding themselves between nursing periods. In those intervals, the pups have to fast for periods that can be quite long. They must have a very rich milk to be able to do this and still grow normally.

Female Elephant seal, nursing.

The two teats are conspicuous.

Contrary to information in several books, a female can nurse several pups.

GROWTH

In their first months of life, calves grow at astonishing rates. Each day, the calf of a Blue whale, for instance, grows 3 to 4 cm (1.2 to 1.6 in.), and its weight increases by 81.5 kg (179.7 lbs.)

The calf certainly works hard for such growth, though. It consumes 90 kg (198 lbs.) of milk daily, the equivalent of more than 370,000 kcal, a fact that should send shivers down the spine of weight-watchers everywhere! As far as assimilation is concerned, these animals must hold the record, being able to convert 90% of the ingested milk to bodily growth and maintenance.

Fin whales produce 72 kg (159 lbs.) of milk daily. The lactation period lasts six to seven months. During that time, a young Blue whale will increase its weight by 17 tons, a Fin whale by 11.5 tons. At this point, they will then measure half their adult size. Such a weight gain is necessary and will be used to provide a food reserve and thermal insulation when the time comes to migrate to the colder summer waters.

A human baby's weight is merely doubled six months after birth. Two more years are needed for the baby to reach half the size of an adult.

In most cases, Mysticetes females fast during the nursing period, which takes place in areas where the water is not productive but at temperatures better for the calf survival.

Physical maturity is attained when the epiphyseal cartilage in the long bones and vertebrae has ossified. Cetaceans as well as all other mammals, thus reach sexual maturity before physical maturity. For instance, puberty in Mysticetes is reached towards the age of six or seven, whereas physical maturity occurs around the fifteenth year of life. Sperm whales reach puberty at the age of 20 and are physically mature at about 35.

Longevity (or life expectancy) in some species of cetaceans can be surprisingly long. Japanese whale hunters have killed, in Antarctica, a female Fin whale 111 years old. It is possible that other big rorquals could live more than 100 years. A new technique, based on aspartic acid racemisation in the lens of the eye gave a maximum age of 211 years for a Bowhead whale. Three others were 135, 159, and 172 years old. The Sperm whale can live more than 80 years, the Pilot whale 63, and the Killer whale more than 45. Smaller species rarely live more than 45 years. Pinnipeds can probably live for more than 35 years, mainly the bigger species.

But how is age determined in cetaceans? In Odontocetes, it is relatively easy. A cross or longitudinal section of a whale's tooth, followed by polishing and coloration treatments, reveals visible dentine rings similar to the growth rings of a tree. One then simply has to count these rings to get an idea of the animal's age. The same method can be used to know the age of pinnipeds. Mysticetes lack teeth, so other and less obvious avenues have to be sought. In their auditory canals one can find an earplug made from the accumulation of desquamated cells and cerumen that has built up through the years. Unfortunately, in addition to the fact that an intact ear plug is hard to get (it has a tendency to melt), no one knows exactly if one or two layers accumulate in a year. Another method consists in counting the bone

Section of a Beluga tooth 17 years old (each line in the previous growth layers group counts for one year. The known life span of the Beluga whale has doubled since this discovery)

layers that apparently accumulate throughout the life of the whale in the tympanic bulla (or auditory bulla), but it is not a very precise one. The best method, that can be only used if the eyes are taken out immediately after death, is based on the racemisation of the aspartic acid contained in the center of the lens of the eye. It was used with success on Bowhead whales freshly killed by Inuit people of Alaska and Minke whales killed by Norwegian whale hunters. It is almost impossible to use this method on stranded whales unless they strand alive, and their eyes are taken immediately after death; otherwise the results are not usable.

CARING FOR THE YOUNG

Cetaceans, at least the larger ones, calve every two years and sometimes every three or four years. Since their life expectancy is not very long, only a few calves are produced.

As with other creatures, care for the young is inversely proportional to the number of descendants produced. Having few offspring usually implies considerable investments in order to ensure their survival. Let us not through anthropomorphism speak of the "love" whales have for their young as a phenomenon particular to these species, or, in other words, invest them with a superior intelligence. They simply follow the reproductive strategy of all species that similarly have a few descendants in every generation. It is not a thought-out behavior, but one that is entirely programmed by their genes.

I saw, some thirty years ago, a female Beluga trying to submerge her floating, probably stillborn, calf. In spite of the nearness of my boat and the possible danger it presented, the whale kept jumping vertically out of the water so as to land on the dead calf and carry it to a certain depth. As the dead whale eventually floated back to the surface, the female would start all over again.

We may deplore the actions of people who would take advantage of this mother's behavior in order to kill it, but it is no better to treat this whale as a *mater dolorosa*, consciously weeping the death of her calf.

Although such an anthropomorphic interpretation is seen more and more frequently in documentaries and books offered to the general public, it should be treated with caution. More will be said later in the chapter on whales and man.

We do not need to give whales human qualities in order to appreciate them. It is sufficient to know that they are magnificent, peaceful, and irreplaceable (as is every species on earth).

PINNIPEDS

Phocids nurse their young for a very short time, but give them a very fat milk. The females protect their young only during this period, then leave and let them fend for themselves. This must be the reason why the pups fast for so long after weaning: they have to learn by themselves how to hunt their prey. Even if the nursing period is short, it is costly for the female. A Northern Elephant seal female weighs 500 kg when she gives birth, and the pup weighs 40 kg. Four weeks later when it is weaned, the mother weighs only 290 kg, but the pup 140 kg.

Often in July along the St. Lawrence young seals, seemingly in bad shape, are found by people walking on the beaches. The good samaritans think they are sick and want to help them by taking them to organizations that are known to care for marine mammals. Most of the time they are freshly weaned pups, not at their best because they do not yet know how to feed themselves properly. In such cases one should never touch or try to feed these animals. They have to learn to feed and fend for themselves. Trying to "help" them will put them in danger: more stress, maybe contamination with germs, etc. Just let them be. You can observe them from afar, but do not interfere with their learning phase; you could do more ill than good.

Otariids, which nurse one year or more, invest more in their young. This investment is costly for the female and is a cause of increased mortality, but it may give a better chance to the young. Some times, a new pup is born while the older one is still nursing. This will increase the chances of mortality of the new pup, because it cannot compete with the older one. Several pups die on the haul-out sites, from wounds caused by adults when they move fast.

Young Hooded seal pup. By its looks one could think it dying, but, as soon as the tide rose, it came back to life and went away.

Young Gray seal pup. Even if it looks very bad, do not touch it! Its mother will come back to nurse it.

PARASITES AND ENEMIES

The large cetaceans, those leviathans of ancient stories, are so enormous that people once believed trees grew on them, as though they were floating islands.

Such stories are a bit exaggerated, but we must admit that these animals are hosts to a myriad of vegetal and animal species. Some of these species simply hitch a ride and do not seem detrimental to whales. They include diatoms (*Cocconeis cetiola*), unicellular algae that color the skin of the host with various shades of yellowish brown. The name "sulphur bottom" that whalers gave to Blue whales comes from the algae on their belly. These often highly visible patterns change every year and so cannot be used to identify individuals from one season to the next.

A small island? No, the carcass of a blue whale on the reef of île d'Anticosti, August 1985.

Diatoms can die or fall off as the whales migrate from the cold waters to their wintering grounds, or vice versa. Among the animals that hitch rides on whales are acorn barnacles (*Coronula sp.*) and stalked barnacles (*Conchoderma sp.*). Swimming larvae attach themselves to the skin of the slower whales in colder waters, as they would to rocks or wharves.

Diatoms on the side of "Bossu," a Fin whale.

Xenobalanus globicibitis
Laurent Soulier

Stalked barnacle (Caribbean Sea).

Once attached, the larva undergoes a complete metamorphosis to become a SESSILE organism, enclosed in a bivalve carapace composed of calcareous shell plates. It settles on the whale's body (in the case of acorn barnacles) or connects itself by means of its muscular peduncle (stalked barnacles). Some Humpback whales can carry up to 450 kg (992 lbs.) of barnacles by the end of summer, but will rid themselves of these hitchhikers in the warmer waters of their wintering grounds. Although their presence slightly increases drag by reducing the whale's streamlining, they do not appear to cause any real damage to their host. One species of fish, remoras (*Echeneis sp.*), can also be carried by the whales, without harm. It would be more appropriate to consider them commensal organisms rather than true parasites.

A parasite depends on its host for food and is, in some ways, detrimental to it. Only in rare cases will parasites actually kill their host. Nonetheless, the fact that they are very numerous can create extremely debilitating effects on the animal. Many whale parasites fit this category. Among these are ectoparasites, which live externally on the animal, and endoparasites, which live inside its body in various organs.

The most visible of the ectoparasites is the crustacean amphipod known as "whale lice" or cyamid. The only characteristics it has in common with true lice, which are insects, are body shape and habits. We find cyamid in all slits, grooves, wounds, around the eyes, blowholes and lip commissures. In a California Grey whale, more than 110,000 whale lice were found in only two wounds.

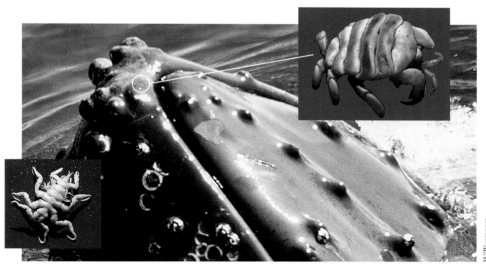

"Whale lice" (*Cyamus sp.*) on the head of a Humpback whale. Similarly, to the left, a whale lice (*Neocyamus physeteri*), commonly found on Sperm whales. Courtesy of J.L. Fabre.

These crustaceans do not have free-swimming stages in their life cycle so contamination is only made through direct contact. In their evolution, cyamids have diversified or adapted to become species specific and even to sub-populations within a species. For example, the Blue whales of the northern hemisphere, isolated for the most part from those of the southern hemisphere, carry different species of cyamids. We use this fact to distinguish whale populations of the same species; if individuals do not carry the same "lice," we may be sure they belong to groups that have never had contact with one another.

Immediately beneath the skin, in the blubber, another type of crustacean is found, a copepod, which starts off as a free-swimming larva but later attaches itself to the skin of the whale. After piercing the skin and penetrating the blubber, it undergoes a complete metamorphosis, with only its egg-carrying abdomen protruding from the skin.

Whale lice and acorn barnacles on a Grey whale.

Drawing of *Penella sp.*

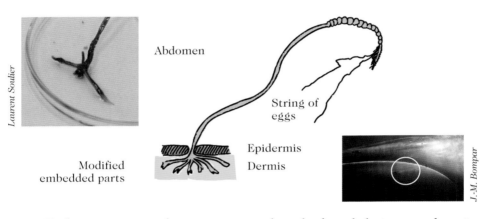

Endoparasites are also numerous and can be found almost anywhere in the various organs. Many species of roundworms (nematodes) inhabit the ears, lungs, and intestines; the Sperm whale's placenta is host to one species, *Placentonema gigantissima*, which can exceed 8 m (26 ft) in length. Other endoparasites include tapeworms, trematodes, and cestodes.

The whale worm, *Anisakis simplex,* in the stomach of a Minke whale.

Tapeworms in the intestines of a Minke whale.

Pinnipeds have their share of parasites too, mostly internal ones, very often the same as those found in cetaceans.

Barnacles can be found some times at the base of the tusks of the Walrus. Pinnipeds can also die massively following infections by germs like Morbilliviruses. In 1988, more than 17,000 Europeans Harbour seals died of an infection similar to canine distemper. They can be victims of phytotoxins associated with red tides, as may have happened to the Mauritanian colony of the Mediterranean Monk seal (*Monachus monachus*) in 1998. Approximately 80% of the colony died, which was a terrible loss for the species that counted only a few hundreds of animals in their whole area of distribution.

We may also find flukes in the liver and aerial passages as well as cestodes in the blubber. The consequences of their presence vary.

Flukes in the liver of a Harbour porpoise.

Cestodes, probably *Phyllobotrium delphini*, in the blubber of a Sperm whale (île d'Anticosti, Québec, June)

Stomach contents of a Grey seal with nematods.

Parasites in the bones of the head of a Harbour porpoise.

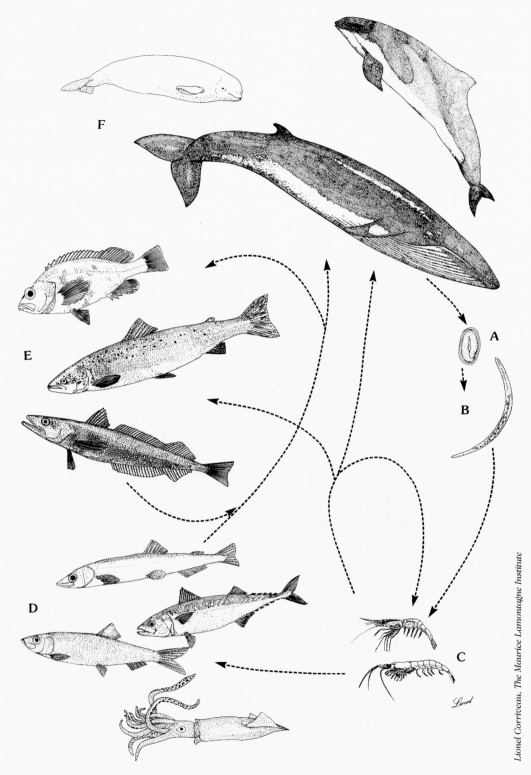

Life cycle of the whale worm, *Anisakis simplex*.

Cetaceans, the larger ones at least, have only a few predators. Some large sharks are known to attack mostly juveniles and calves. On the body of cetaceans, round scars with clear, well-defined edges can be seen. A small warm-water shark (60 cm; 2 ft), known as the cookie-cutter shark, or its larger cousin of the genus *Istiophorus* probably made these wounds. It swims toward a whale and bites, obtaining a firm grip. As the whale swims, the shark is slowly turned around, cutting out a deep circle of flesh (epidermis and blubber) from its prey.

Isistius brasiliensis, the cookie-cutter shark.

Drawing of a *cookie-cutter* shark, *Isistius brasiliensis*.

The smaller species have to deal with sharks, polar bears, and other cetaceans, all dangerous. The Killer whale can attack practically anything that swims, other cetaceans included. Near the Mingan Islands in the St. Lawrence River, Richard Sears of the MICS (Mingan Islands Cetacean Study) has witnessed at least one such encounter between Killer whales and a Minke whale, an attack of lethal consequences for the

Minke. These super-predators hunt in packs or pods, much like wolves, and will even take on a blue whale. It seems likely that the Killer whale can attack any species it encounters. We have found the rostrum of a swordfish in the body of certain whales. This is probably a sign of the fish's aggressive behavior rather than attempts of predation!

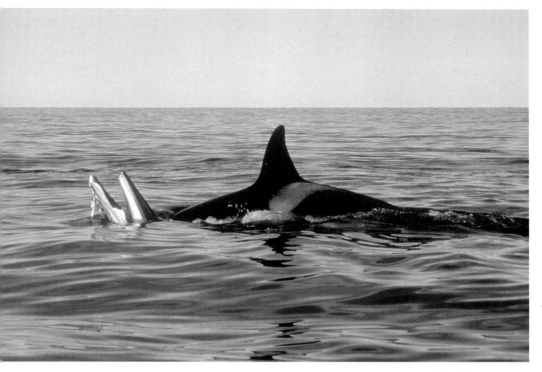

Minke whale attacked by Killer whales near the Mingan Islands (Québec).

Richard Sears

Young Greenland seal pup may have been killed by a shark.

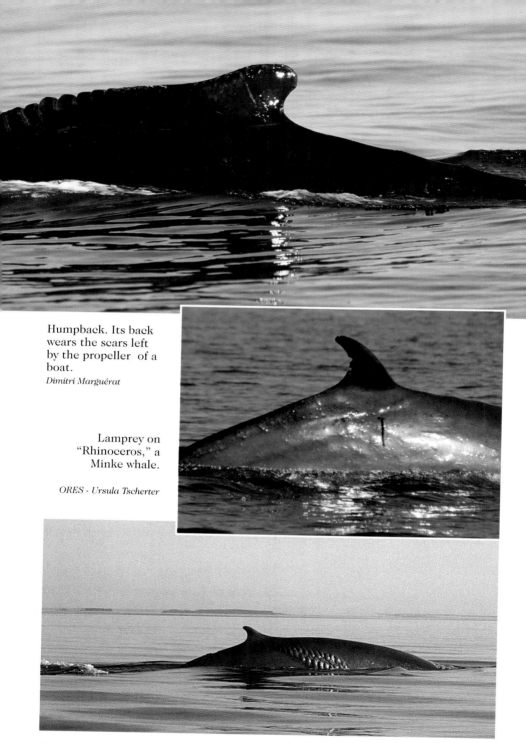

Humpback. Its back wears the scars left by the propeller of a boat.
Dimitri Marguérat

Lamprey on "Rhinoceros," a Minke whale.

ORES - Ursula Tscherter

"Zipper," a Fin whale bearing the scars left by a boat propeller.

WHALES AND HUMANS
HUNTING OR "FISHING"

Oddly enough, governments usually call on their department of fisheries to regulate the exploitation and protection of cetaceans. Is this practice a reminder of the days when cetaceans were considered fish so that Christians could eat their flesh on Fridays, a day when meat was forbidden? Or why do people in some countries continue to "fish" marine mammals, but do so with rifles as their "tackle?"

Our associations with whales undoubtedly date back to our very beginnings as a species. Archeological sites have revealed tools made of stone and whale bones. In the north of Norway, rock engravings of cetaceans, which sometimes included fine details about their anatomy, have been found. Small cetaceans such as dolphins or porpoises are usually represented in these carvings. The fact that we often see engravings of whales in the company of elks suggests that they too were hunted. Small kayak-like boats included in these representations support this conclusion.

Neolithic wall drawing found in *Man and Whales* by R. Ellis p. 91.

Stone tools found alongside the bones of the great whales probably bear witness to an opportunistic use of whale carcasses rather than one resulting from a coordinated hunt (although some groups could have hunted whales). A South Korean wall engraving shows a great whale being harpooned and towed by what may be a huge canoe. Although difficult to date, the engraving could be more than 6,000 years old and would seem to provide information concerning the whaling techniques used during the Neolithic period. Whales were used for various purposes. Some people were mostly interested in the blubber and, occasionally, the skin (when it was thick enough, as in the Beluga or the Narwhal, for instance). Others used the entire animal, feeding on its flesh and blubber and using its bones as raw material.

South Korean wall drawing of a whale hunt involving a large Mysticete.

The slower and more coastal species (e.g., the Right whales and the Grey whales) that usually ventured into the shallower waters were probably easier to hunt. The aboriginal people of Siberia and the Canadian Pacific coast as well as the American Inuits (or the Eskimos) were already hunting whales when the first Europeans arrived. The Eskimos mostly used one-man kayaks, while the Amerindians, in what is today British Columbia, used large eight-man pirogues or dugouts. Their equipment and techniques were probably very similar to the ones used by the Neolithic hunters. As for the Japanese, they had already hunted the Right whales, and even the rorquals, for that matter, in the Sea of Japan. These whalers used huge nets to trap even the faster whales. Once a whale was caught, a fleet of whaleboats would surround it and killed it by plunging lances into its body.

West coast Amerindian dugout, American Museum of Natural History, New York.

THE BASQUES AND WHALE HUNTING

Whaling, then, could have been expected anywhere human beings encountered whales. The Basques were the first, however, to make whaling a livelihood and to commercialize it in the occidental world. The first documents to report such activities date back to around the year 1000 AD (records from Bayonne Cathedral). One could even think that they were hunting whales in the VIIth century since they furnished oil to Jumièges Abbey then. They hunted the Biscay Right whale (so called because whaling had initially been carried out in the Bay of Biscay), known as *Balaena glacialis* (the Black Right whale). They probably also hunted the last of the Atlantic Grey whales, contributing to their extermination. No one knows exactly when the last Grey whales disappeared from the Atlantic, but accounts of *Sandloegia* in documents dating back to 1640, descriptions of the *scrag whale* by Paul Dudley, Esq., in 1725, and those of an animal called *otta sotta*, in 1611, by the directors of the Moscovy Company are considered by Mead and Mitchell (1984) reliable proofs of the relatively recent presence of the Grey whales in the Atlantic. In the beginning, the Basques may have herded cetaceans into the shallower bays where they were lanced to death. Still practised today in the Faeroe Islands, this type of hunt can be very effective: it was reported that a single drive was responsible for the capture of 1,540 Pilot whales in the Shetland Islands on September 22, 1845.

However, the shores of their country were not really appropriate for hunting this way. Though it is said that they learned to hunt from Viking immigrants in northern France, it is more likely that, very early, they invented their own methods and hunted along their coasts with boats, since the whales did not pass very far from it during their seasonal migration. As coastal whales became scarce, the Basques took to pelagic or deep-sea whaling. One does not know if it was the coastal hunt that made the cetaceans scarcer and made it necessary to begin deep sea hunting. It may have been changing climatic conditions influencing the distribution of their prey, or more simply that the whales, long living animal as they are, had learned to shun the areas where they were harassed by hunters and stay farther from the coast.

An attalaya or watch tower, used in whale hunting, near Ghétarry, Basque country. The picture has been edited to remove a shack which disfigures it. On the right the attalaya as seen from the sea.

The Basques were hardy navigators. Because their land resources were poor, very early they turned towards the sea to find a more decent living. Maybe it was the tales told by cod fishermen that induced them to go to the "terres Neuves" to hunt whales, which were abundant there. We have records that show that there were cod fishermen very early in the XVIth century in the Newfoundland area.

Since historians have begun to search archives from the Basque country (Laurier Turgeon, archeologist from Québec, did so in France and Selma Barkham in Spain). Hundreds of documents, acts, contracts, drafts of law suits, etc., have been found, that shed new light on the importance of maritime traffic between the Basque country and America.

Biarritz city coat of arms, Basque country.

In 1571, the Basque fleet counted more than 100 vessels, of which 25 to 30 were whale ships, that gave work to more than 4,000 sailors and fishermen. In 1578 one could find between 30 and 40 whaleships around Newfoundland. In searching those archives, Canadian archeologist Selma Barkham learned about the wrecking of a Basque galleon, the San Juan, in a harbour which she identified as Red Bay on the Labrador coast. The documents she studied incited her to mount an exploratory expedition on the Labrador coast, where she, her son Michael, and other archeologists, identified several sites where Basque whalers had established hunting settlements. Archeological digs were done at Red Bay, on Saddle Island, and in the bay between the island and the coast.

Red Bay as seen from Saddle Island.

Red Bay, called at the time Buytres or Boytus, seems to have been one of the places preferred by Basque whale hunters, from the first half of XVIth century to around 1620. Located in a bay sheltered by an island (Saddle Island), on the north shore of Belle-Isle straight, where migrating whales passed during their migrations, it made the hunting of the whales with small boats easier.

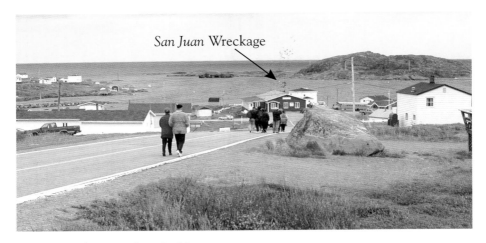

Saddle Island as seen from Red Bay.

The digs uncovered the vestiges of buildings and tryworks, and the very well preserved remains of the wreck of a galleon, possibly the San Juan which had been owned by Ramos de Arrieta said Borda from Passajes in the south part of the Basque country. As cold water and the accumulated silt had preserved it well, it was possible to learn quite well how a galleon was built at the time. Under the galleon were the wrecks of two "chalupas," the smaller boats used to hunt whales. Those wrecks are for the time being, the second oldest wrecks known in North America. Land digs on Saddle Island gave lots of information on how a whaling station was built in those days.

Model of the whaling ship "San Juan," Red Bay Museum.

A "chalupa," restored wreck, Red Bay Museum.

A crew of 75 men was needed to man a 250-ton burden whaling ship like the San Juan. It would have carried 1000 barrels of oil (each barrel contained around 211 litres). Among the crew were sailors, for sure, but also specialized workers, like carpenters and coopers. Carpenters maintained the ship and built the sheds and houses needed for the work on shore. Coopers maintained the barrels that had been brought whole, filled with the water and food for the trip in good shape and assembled those that had been brought in pieces to save space. The coopers lived on land, the rest of the crew stayed on the boat, which had been anchored and made fast to the shore in the safest place possible (which was not always secure, based on what we can infer from the case of the San Juan, which was thrown and wrecked on the shore of Saddle Island in a storm!) The living was very hard, both during the trips and on land. Part of the crew did the hunting. Very often crews from several whalers cooperated to capture whales, drawing lots for their share of the bigger catches.

Cooper, Red Bay Museum.

199

Model of the tryworks found on Saddle Island. Red Bay Museum.

Once killed, the whales were towed to the ship in the harbour, where they were flensed. (Sometimes they were pulled onto land). Blubber was then brought to land near the tryworks, cut in smaller pieces and rendered. Carcasses were pulled at sea. Fire in the tryworks were started with wood but soon enough there was enough residue in the pots to maintain the fire underneath. The oil was regularly ladled, cooled, cooked again if necessary, and then put in the barrels. Finally they were loaded onto the boats. All this needed constant work, the more so if the hunt was good. The smell around the place would have been awful because lots of carcasses must have washed ashore and rotted there. In 1705, Augustin Le Gardeur de Tilly, Lord Courtemanche, wrote, speaking of the then abandoned whaling station: "one can still see the tryworks and bones of whales... We counted 90 gigantic heads in only one place." Even now one can find plenty of whale bones lying around the village of Red Bay. One can imagine the stench when they were still meaty!

Ruins of the tryworks Saddle Island, Red Bay.

Basque whaler, Red Bay Museum.

Some years, when the hunt had not been good and there was still room in the holds of the whalers, the captain and even the crew (who were paid in proportion of the cargo brought back), may have decided to remain longer than it was safe to do and were caught by the ice. In some cases they could even choose to stay for the winter. If it was particularly harsh, the consequences could be dramatic. Records mention that during the winter of 1577, 540 people died in Red Bay. Archeologists found the skeletons of 125 people on Saddle Island, maybe some of those poor guys. Their death was not completely wasted because local conditions have favored the conservation of many pieces of their clothing, giving us good informations about the way a Basque sailor dressed. It was in Red Bay too that the second oldest will known from Canadian history was written, the one from Juan Martinez, from Larrune, June 22nd 1577.

Cutting marks left on a vertebra that may have been used as a cutting block.

Part of a whale skull on the shore. Saddle Island is in the background.

Towards the end of XVIth century the hunting of whales moved more to the west, in the Saint Lawrence, where archeologists have found other ruins of Basque whaling stations in Grandes-Bergeronnes near Tadoussac and across the river, on île aux Basques. The ruins date to the late sixteen century.

Shoe, harpoons, flensing knife, Red Bay Museum.

Remains of Basque tryworks on île aux Basques, near Trois-Pistoles (Québec).

Well-preserved remains of tryworks used to melt whale blubber were also found on that island. The archaeologist Laurier Turgeon and his team have uncovered traces of a settlement, a collection of fragmented tiles, and the tip of a right whale's mandible, perhaps one of the bones that Father Nouvel mentions in one of his accounts of a visit of an abandoned Basque settlement.

"This island... is most pleasant.... It is called île aux Basques (Basques Island) because the Basques carried out whaling activities on it. I enjoyed visiting the tryworks where they used to produce whale oil; we can still see the long ribs of the whales they killed."

Tip of a right whale's mandible found on île aux Basques by archeologist Laurier Turgeon.

It should be mentioned, though, that the fur trade, not whale hunting, may have been the main reason for creating these settlements,. (Michael Barkham).

The Basques ceased hunting whales at the end of XVIIth century, probably for economic reasons (European conjuncture, lack of raw materials to build ships and to make iron due to too much tree cutting for example).

It is very impressive that so small a country could have had so much importance on maritime life of the time. The Basque navigators had shown the way, leading to the exploration of the whole world, and, for a long time, Basque navigators and pilots were employed on many ships, transmitting the experience gained during their heyday.

As long as it endured, whale hunting was very lucrative for the Basques. One whaling ship could bring back a cargo of 1,000 to 1,500 barrels of oil (each one worth 6 to 8 ducats), to which was added baleen (whalebone), which was worth as much as oil in the second half of XVIIth century. According to Pierre Proulx who made a comparison between the salary of a carpenter (35 ducats) of the time and a modern carpenter salary ($25,000), the total value of the 1,000 barrels of a ship like the San Juan had a value in excess of $5 million of today's money. They were the oil tankers of the past! One has to add to those direct revenues all the connected ones: ship building, armament, and so on.

AFTER THE BASQUES

The British and the Dutch soon took over whaling in North America. For a long time, however, crew members, particularly harpooners, consisted of Basques. The Americans soon joined the whaling vessels that hunted the Right whales from Spitzberg to Davis Strait, between Greenland and Labrador, in the Hudson Bay and beyond, toward Baffin Island. They hunted the Arctic Right whale to near-total extinction. In 1913, the two remaining whaling vessels to hunt in these waters, the *Morning* and the *Balaena* returned from a fruitless expedition to their British harbour. Right whaling in the Greenland waters had come to an end.

After having exterminated the north Atlantic Right whales, whalers concentrated their efforts in the north Pacific, between Asia and North America and, simultaneously, in the southern hemisphere: around Australia, New Zealand, and South America during the austral winter, and in Antarctica during the summer. As whaling intensified, it again caused significant drops in the numbers of Right whales. Whalers ultimately had to turn to other species, which, until then, had been but sporadically hunted, either because of their great speed or because they inconveniently sank to the bottom once dead.

The Americans are considered to be the first on record to hunt the Sperm whales. Captain Christopher Hussey killed the first one near Nantucket, in 1712. However, it is worth pointing out that an archeological excavation in Spain has unearthed the bones of Sperm whales in a Basque settlement dating back to well before the sixteenth century. Many documents mention that toothed whales, from the head of which several barrels of oil could be extracted, were hunted by Basque whalers. These products were treated in Biarritz and Saint Jean de Luz early in the XVIth century.

E. de Cleirac in his book published in 1671 (*Les Us et Coutumes de la Mer*, quoted by Nelson Cazeils) mention such hunting:

> *"There is a species of whale, which have small flat teeth and no whalebone (baleen) in the mouth, and from this one, Basques extract the product named Whale sperm, which is nothing but the brain of the beast, with which they fill barrels drawn from the head with pots and spoons. Druggists work it, wash it, and prepare it so it becomes white as snow or flower of salt, smelling like violet, surpassing any whiteness. This is used to make excellent make-up and they have called it sperm for the sake of the women and the girls that use it."*

This shows without any doubt that that the Basque whalers did hunt the Sperm whale before Americans did it.

It seems that nothing could stop the Basques. Furthermore, it is fair to say that they certainly did break new ground in all fields of whaling. As far as the odor of violets of the product, it must have have been after a serious treatment, because Nelson Cazeils quotes the magistrates of the city of Bayonne who had to forbid the rendering of "the brains of

Model of tryworks on board a whaling ship. New Bedford Whaling Museum.

Sperm whales" in the city limits.

> *"...But there are some peoples who dabble in the trade of spermaceti. ...Particularly the man de Lure, who is now occupied with the rendering of sperm whales brains in a house situated in rue de La Saillie... the stench from this rendering is spilling into the whole neighborhood, and is so awful that all the neighbors are suffering from a great nuisance in their houses, with a continuous infection that make them almost uninhabitable, from what they have complained to the Magistrates..."*

All those who have been near a dead whale can only be very sympathetic with the poor neighbors!

The Americans hunted Sperm whales on vessels reminiscent of the Basque chalupas. They were more solidly built, however, allowing whalers to attach the harpoon line to the vessel itself, instead of the float that the chalupa tried as best it could not to lose. As the wounded animal tried to escape, it would sometimes pull the whaleboat for hours. These episodes were referred to as "Nantucket sleigh rides" and were popularized by Melville in *Moby Dick*. On such occasions, the harpooned animal would obviously tire faster. The great rorquals that were too rapid to pursue and that sank when killed were, at first, spared by whalers. The invention of a harpoon equipped with an explosive head, which was fired by a high-calibre cannon, soon ended the relative quietude of these whales. Whalers had long been able to fire harpoons from cannons, but such cannons were more like large rifles and fired only standard harpoons, which were sometimes little larger than hand harpoons.

The "Spes et Fides," one of the first of the modern whale ships, launched in 1864. One can see the cannon and the exploding head harpoon.

Reproduced from "La Baleine," translated from English by Roland Beauchot, Paris. Stock, 1972, p. 188-189.

Harpoon gun and harpoons, American Museum of Natural History, New York.

After many years of research, a Norwegian named Svend Foyn developed a large gun capable of firing a harpoon armed with an explosive head. This new weapon would practically annihilate all of the great cetaceans. Foyn's system was patented in 1872. The impressive calibre of the cannon made it possible to fire heavy grenade harpoons considerable distances. Furthermore, the development of steam-driven vessels to replace the slower sailing ships allowed for the attachment of the cannon to their bow. Consequently, whales that sank were no longer a problem since these vessels could now hold the carcass long enough to inflate it with air and keep it afloat.

New whaling stations appeared in all the oceans of the world, particularly in the southern hemisphere where whales were killed by the thousands. A man named Peter Sørlle patented the stern slipway, which consisted of a ramp that began just above the waterline and enabled whaleships to winch the whale aboard where flensing could then take place. As a result, the first floating factories were created. Until then, whales had to be flensed alongside the boat, a real tour de force in bad weather, or hauled to shore-whaling stations. Freed from these constraints and manned with expeditious whalers, the floating factories now could hunt whales anywhere. This obviously brought on an even greater carnage of the already decimated herds of large cetaceans. The smaller species (Bryde's, Sei and Minke whales) were also soon slaughtered.

In 1939, prior to World War II, the 34 floating factories in the Antarctic had killed 36,000 whales and brought back 2.7 million barrels of oil. As early as 1931, people had realized that existing stocks could not sustain such a rate of destruction. The whaling industry, motivated more by a desire to maintain existing prices than by ecological concerns, tried to limit oil production, but without much success. Quotas were established, but some of the more directly concerned countries refused to sign the *International Convention for the Regulation of Whaling* established by the League of Nations in 1931. The war provided only a temporary respite for whales but, in 1948, 44,000 whales worldwide were killed. From 1953 to 1966, more than 50,000 were killed annually, reaching peaks of 60,000 from 1958 to 1965 (66,090 in 1966 is the all-time record of registered catches). Afterwards, catches began to systematically decline, falling below 2,000 in 1982-83. This low number may be partly due to efforts by the whaling industry to reduce the catch, but it reflects just as much the degree to which whale populations had become endangered.

Bridge and interior installation of the factory whale ship.

1. Ramp to hoist whales on board
2. Recreation room
3. Hospital
4. Crew quarters
5. Engine room
6. Boilers
7. Laboratory
8. Flensing deck
9. Trypots for the blubber
10. Acess panel for the blubber grinders
11. Carpenter's workshop
12. Machinist's workshop
13. Fishing winches
14. Rotative blubber trypot
15. Oil separator
16. Liver conditioning
17. Smithy
18. Bone saw
19. Bones trypots
20. Fuel tanks
21. Cutting deck
22. Oil separator
23. Meat rotative trypot
24. Machine to fill the meat bags and conveying belt through the plant
25. Pump room
26. Lift to the loading platform
27. Radio room
28. Bridge and annexes
29. Crew dning room and kitchen
30. Freezers
31. Food magasin
32. Bosun store
33. Food storage room

208

Despite the repeated warnings of biologists throughout the world, it was only in 1982 that the International Whaling Commission (IWC, created in 1946) decided, by a vote of 27 in favor, 7 against, and 5 abstentions, to abolish all commercial whaling by 1986. Japan and the Soviet Union were against this resolution. Canada, although a whaling nation, had passed a moratorium on whaling in its waters in 1972, and withdrew from the IWC in 1981. In any event, Canadian whaling had already been declining before that time. For example, the Newfoundland Whaling Company, created in 1900, had established 18 whaling stations by 1905. In 1950, only 3 remained. Canada's withdrawal from the IWC served only to reinforce its disinterest concerning the trade of whale by-products.

Harpoon cannon, Oceanographic Museum of Monaco.

Commercial whaling is still not completely banned. Japan, Norway, and Iceland continue to hunt the Minke whale for its meat and blubber. Other types of whaling, including traditional and scientific whaling, also take place. To justify its harvest of Minke whales, Japan announced, on May 10, 1996, a first in the field of *in vitro* fertilization: scientists had managed to fertilize Minke whale ova (which matured *in vitro*) with frozen spermatozoids. Their goal, or so they say: comparative research for the protection and betterment of endangered species! Yutaka Fukui, professor at the University of Veterinary Medicine and Agriculture of Obihiro, has admitted, however, that scientists interested in artificially increasing cetacean populations will have to solve the not so insignificant problem of placing the embryo in the body of a female!

Since this hunt mainly involves the Minke whale, a species not yet considered endangered, we can hope, in spite of everything, that the fate of these magnificent animals will continue to improve, as long as pollution does not contribute to the death toll. In fact, except for the case of certain species of porpoises (Vaquita) and some fresh water dolphins there seem to be no remaining species on the verge of extinction (even if we consider the case of the Northern Right whale, and admit that the northern and southern hemisphere populations are subspecies (borealis and australis) of the same species, *Eubalaena glacialis*, which is put to the question nowadays.

Most species have benefited from protection. The Grey whale population in the eastern Pacific, about 15,000 to 20,000 before it was hunted, had dropped to less than 2,000 by the time the ban was imposed. Since then, this population increased to 21,000 and it was taken off the endangered species list. The herd of Humpback whales that winters in the Australian waters has increased from 200 to 2,800 and probably even more (see the chapter on migration). Some feel that the Minke whales in the southern hemisphere are more numerous than they have ever been. They apparently benefited from the high abundance of krill, a result of the great reduction in the numbers of Blue and Right whales, two species that are great krill consumers. The most endangered species is the Northern Right whale. A maximum of 1,700 Black Right whales in the northern hemisphere (less than 350 in northern Atlantic), and 8,200 Greenland Right whales or Bowheads remain. They face an uncertain future. Only the aboriginal peoples have the right to hunt these whales. The danger to the Right whale may thus seem minimal. Nonetheless, any number of kills, as far as an endangered species is concerned, may be excessive, especially when we consider the current hunting techniques. Many whales that are killed, not by traditional methods such as harpooning, but by rifle, cannot be salvaged since they sink to the bottom or disappear under the ice flow. The Southern Right whale population exceeds 8,000 animals, and is steadily growing (10% a year). It does not seem in danger any more.

Some nations are eager take up whaling again, at least of species whose populations can sustain it, such as the Minke whale. There is nothing wrong with killing a few individuals as long as it is done intelligently and by respecting the reproductive capacity of the species. It improves population dynamics, lowering, among other things, the age of sexual maturity, and stimulating reproduction. This likely holds true for cetaceans as well. But again, are human beings able to adequately manage the exploitation of whales? The idea of maximum profit in a minimum of time is the driving force behind our capitalistic societies (and all others, whatever their disguise). Will a country that is again officially given the right to practice whaling respect the imposed quotas? As long as ideas and philosophies remain unchanged, it is better that we respect the current moratorium—which, as we have seen earlier, is far from being unanimous since many countries and aboriginal peoples continue the hunt for various reasons. Not until everyone has learned that it is possible to use and not abuse will we be able to appropriately exploit the richness of nature's resources.

POLLUTION

Pollution is one of the most treacherous factors in slowing down the rate at which cetacean stocks are renewed. The threat to the population of Belugas, or white whales, in the St. Lawrence River, illustrates this danger. The Belugas seem incapable of substantially increasing their numbers in spite of protection. The reason? Belugas feed not only on pelagic fish but also on benthic animals, that is, organisms living on the bottom and/or in the marine sediments, particularly the marine polychaetes of the genus *Nereis*, or clam worms. Bottom sediments, from which the polychaetes are extracted, are found in the whales' stomachs (Vladykov, 1946) and contain a variety of pollutants: heavy metals, organochlorines, Mirex, PAHs, etc. Many such contaminants can accumulate in high levels within the animal's body and especially in its fat. When food becomes scarce, the animal draws from its fat, thereby circulating these toxins through its body. We have seen that cetacean milk has a high fat content. Part of the contaminated fat is used to produce the calf's milk needed to ensure growth and survival in its native cold waters. Consequently, the baby whale is contaminated at the nipple; not a very good beginning! But this inauspicious fate is not restricted to merely the young Belugas. Other cetaceans as well, temporarily protect themselves against these toxins by storing them in their body fat, and will likewise contaminate their nursing young. The Beluga differs from the other cetaceans with respect to the degree of contamination; it is high enough in the Beluga that its carcass is considered a hazardous waste and must be disposed of accordingly.

The greater number of other cetaceans consume less contaminated prey and accumulate smaller quantities of pollutants in their body. But, as small as these amounts may seem, they are still too much, especially when we realize that the newborns are those most affected by pollution. Also, the effects of pollution will only become visible many years from now, perhaps at a time when we believe that we have brought it under control.

When we think of all the nuclear waste that is slowly contaminating the bottom of the oceans, we have every reason to worry about what lies ahead for whales (not to mention for us). Its effects on the environment will last thousands of years. With the few available tools to counteract them, not much can or will be done, unless a technological breakthrough takes place, which, for now, is more in the realm of science fiction. The effects of nuclear radiation are mutagenic. Judging from the aftermath of the atomic bomb attacks on Hiroshima and Nagasaki in 1945, where the irradiated people are still suffering, or that of the Chernobyl disaster and the subsequent cancer rate among the local population, we have every reason to be pessimistic about the fate of all organisms that come into contact with nuclear waste.

During an interview, Pierre Béland of the St. Lawrence Ecotoxicological Institute said something that profoundly moved me: the Belugas are being killed by their environment, but it is our environment as well. Are we next in line?

To all these dangers, we must add one more, which has recently been increasing in importance: the reduction of the ozone layer, the earth's natural sunscreen against the ultraviolet rays (UV) of the sun. Scientists believe that the rate of skin cancer will increase in humans. What effects will the mutagenic nature of UV have on the earth's flora, the basis of all trophic levels? What of the phytoplankton and all creatures that depend on the 80% of the atmospheric oxygen it produces? Finally there is the global warming situation, due, at least in part, to the accumulation of green house gases in the atmosphere. This global warming will have strong effects on climate in general, and the way water circulates in the oceans, affecting the direction and the intensity of the flow of oceanic currents, with all the sad consequences on the upwelling areas. Will the cetaceans and marine life in general be able to change their habits quickly enough to cope with these changes? The time has come for humans to wake up and to take on the only problem critical to us all: the rational management and conservation of the environment that will allow animals, including ourselves, to survive and prosper. What use is there in redressing the economy if cancer kills us all? Dying rich in these circumstances is certainly not more acceptable than dying poor!

ACCIDENTAL CATCHES

Another element must be included among those that can interfere with the recovery of some endangered species. This problem may even represent a threat to populations that have been sparingly or never hunted: accidental catches in fishing gears.

Pierre-Michel Fontaine

Harbour porpoise that has drowned in a fishing net.
The mesh of the net has left visible markings.

Nets have always represented a problem for cetaceans, especially now with the advent of nylon or other similar types of fibres. Because these nets have an acoustic impedance very close to that of water, whales are practically unable to detect them. Huge drifting nets, cod traps, moored gill nets, and the ropes used to haul up lobster or crab traps can also entrap cetaceans. Through their evolution, these animals have never "learned" to avoid such dangers, and those caught seldom live to profit from the experience since they drown (suffocate) in their attempts to free themselves. Only the larger whales may be able to escape. How natural selection could have developed an innate avoidance behavior in whales to fishing nets is hard to imagine.

Once again, the survival of cetaceans largely depends on measures that must take into account regional economic factors, in this case fisheries, which are an essential source of income to local populations. Solutions must be found to minimize the harmful effects fishing gear has on whales, but without denying the rights of those whose livelihood is at stake. Accidental catches will probably never be completely avoided, but ecologists (genuine ones, not animal rights extremists), fishermen, and governments should work together to find viable economic solutions to this major problem.

Fin whale stranded on the shore of the Gaspé peninsula. Its caudal fin is entangled by parts of a fishing net.

Minke whale caught up in the ropes of a lobster cage. Notice the notch made by the ropes on the side of the mandible.

WHALES AS RESOURCES

What were whale by-products used for? Most aboriginal peoples used the entire whale: the meat, the blubber, the baleen (whalebone), the teeth, and, sometimes, even the skeleton. The meat and the blubber were used for food, lighting and heating. The baleen provided elastic material for weapons, brushes, and brooms. The teeth were useful as weapons, tools, and decorative ornaments. American sailors found even another use. By means of the scrimshaw technique (carvings highlighted with lampblack or India ink), seamen would carve scenes on the teeth and bones of Sperm whales depicting their life on board whaling vessels, landscapes, or even the yearning for their loved ones.

Objects in bone and ivory, Whaling Museum New Bedford.

This bone in a Sperm whale jaw did rest Now tis intended for a woman's breast This my love I do intend For you to wear and not to lend.

Bone and ivory artifacts, Whaling Museum, New Bedford.

Model of a ship Ivory, Whaling Museum, New Bedford

Carved Sperm whale tooth, Whaling Museum, New Bedford.

Europeans used whale oil mostly for lighting and in the textile industry. The blubber was also highly appreciated. It was known as "Lenten fat" and could be eaten along with fish on meatless days (days when good Christians could not eat meat for fear of grave sin) even though whales had long been known to be mammals.

Sperm whale tooth, with scrimshaw scene of Right whale hunting.

The baleen rapidly became extremely valuable since it provided a first-rate elastic material. It was put to all sorts of uses: umbrella ribs, corset stays, whips, brushes, hats and carriage springs. Apparently, it was even used to make springs for watches. The French inventors Launey and Bienvenu presented the model of a helicopter propelled by a spring made from the baleen at the French Academy of Sciences, in 1784.

Stays for shirt collars made from baleen.
Peabody Museum, Salem, Massachusetts.

Umbrella at the Peabody Museum.
The ribs are made from baleen plates.

Indicative of the market value of a baleen cargo, the first American steam whaler, the Mary and Helen, which had cost $65,000 to build, came back from her maiden voyage with 2,350 oil barrels and 45,000 pounds of baleen (20,385 kg), all worth more than $100,000. At the time, the baleen was sold for three dollars a pound (0.453 kg). Skilled labourers would earn 90 cents daily, whereas crew members, on average, earned much less, about 26 cents a day.

Sperm oil, or spermaceti, is a lubricant that resists pressure and heat very well. Until as late as 1972, General Motors used it in certain types of differentials (positraction). Apparently, some engines on space shuttles today are lubricated by sperm oil!

Modern uses for whale oil, Whaling Museum, New Bedford.

In Japan, Norway, and in the Faeroe Islands, whales are mostly hunted for their meat, which is extremely expensive—many hundreds of dollars per kilogram in Japan.

Can cetaceans represent more than just meat, bones, and oil to humans? Is it really necessary for us to exploit them this way or are there other conceivable alternatives?

A more "humane" way to exploit whales exists in Canada and elsewhere in the world, and is growing rapidly. Whalewatching occurs in about 40 countries and represents a significant, viable, and profitable alternative to whaling. Whalewatching trips may be done on different types of boats offering varying degrees of comfort. On board these ships, tourists are given the opportunity to see marine mammals in their natural habitat. The boats generally leave from ports that are close to scenic routes where whales are known to gather. Naturalists onboard provide educational information about the trip and about the whales they are likely to encounter, including their biology and ecology. Among the various whale-watching tours are those on large excursion vessels built to carry many people comfortably. In addition, they are usually many naturalists on board equipped with the appropriate educational materials.

Large whale-watching boat.

Other outings focus more on adventure, by offering trips on lightweight, fast-moving, inflatable rafts where less comfort is compensated by a closer, more intimate contact with the sea and its peaceful inhabitants. The touring company may also provide educational information.

Rigid-hull inflatable raft used for whale watching.

The popularity of whale watching has brought on a substantial increase in the number of touring enterprises and has raised questions concerning the extent to which this kind of ecotourism affects whales. The not-uncommon sight of forty-odd ships gathered around whales clearly raises questions regarding their well-being. Not all conditions are appropriate to whale watching. We must not forget that some areas where cetaceans are observed may be used for feeding or breeding. Furthermore, we should also keep in mind that despite their large size and apparent peacefulness, they are nonetheless wild animals, and like all wild animals, may well be afraid of humans. A sudden approach or a prolonged chase can seriously affect their life cycle.

A few precautions must then be taken when approaching a whale: stay at a safe distance from the whale, especially if approaching from behind or from head-on. It is best to approach a whale from the side. Also, most countries that permit whalewatching have set rules regarding the distance to which they may be approached. Some whales, such as the Humpback whale, are known for their spectacular behavioral displays out of the water (breaching). On such occasions, whales have landed on whalewatching boats killing or wounding their passengers. This type of accident is not the result of malice, but rather the whale's inability to detect the boat. Their biosonar, if indeed present in Mysticetes, certainly does not allow them to detect an object floating on the surface of the water, especially if it has a small draft, lacks a motor, or has stopped. It is important to keep the motor running at all times (idling in neutral if not moving). The sound of the motor allows the whale to detect the presence of a boat. If on a sailboat or sea kayak, turn on an echosounder, place a radio against the hull, or strike the hull at regular intervals to signal your presence. I once saw a kayaker get the scare of his life—even if afterwards, he thought it had been the most wonderful experience of his life!—when a Minke whale lifted his boat

Comfort and adventure are combined in this average-size, fast-moving boat.

as it surfaced to breathe. In 1960, at Sainte-Catherine Bay, when I was beginning to learn about whales, a Minke whale breached 2 metres (7 ft) from my boat. And if it had breached 2 metres closer to me? A book would surely be looking for an author!

If our presence brings a change in the whales' behavior, it probably means that we are too close and should move away. Some whales will occasionally approach boats. Caution should be exercised to avoid injuring a whale with the rotary blades of the motor. Letting a whale approach a boat is perfectly safe, as long as it has seen it. They are gentle giants that normally never show signs of aggressiveness.

Some species, such as the Beluga (or the white whale) of the St. Lawrence River in Québec, have a special status. This species is in fact considered close to extinction. No approach maneuvers are permitted since they would subject this population to even more danger, for example, by disrupting the cohesion of the group or separating a mother from her calf.

Some people believe that strict regulations for whale watching should be implemented. Others go as far as to suggest that tours should be abolished altogether. I believe that a heightened public awareness of these creatures' great beauty is the key to ensuring their ultimate survival. If more people go on these tours, more will appreciate the most powerful and most gentle of animals. What a terrible and incalculable loss it would be if these leviathans were to disappear forever. Furthermore, it has never been shown that whalewatching boats, when operated by responsible captains concerned with the conservation of this resource, actually harass cetaceans. In Southern California where grey whales winter, there are now many enterprises that take tourists on whalewatching tours. A few years ago, these whales would flee the instant a boat appeared. Nowadays, they rub themselves on the inflatable rafts and will even let themselves be petted by tourists.

Beluga.

I have noticed a similar phenomenon in the St. Lawrence River, near Tadoussac, an area more familiar to me. Cetaceans (especially the whales that come back every year) seem more and more tolerant toward whale-watching boats. None, as of yet, will let themselves be petted, but their behavior suggests that they are becoming more and more accustomed to touring ships with every passing year. A biologist who tags rorquals using a radio-tag mounted on a suction cup (this is done in order to track whales during their movements and dives) has told me that it became easier to tag them as the season progressed. It is the small craft owners whose principal interest is to get as close as possible to whales that are most likely disturb them. The large and rapid cargo ships that do not slow down also represent a source of danger, especially in areas where there are already many whalewatching boats and where whales are less timorous.

Most of all, it is the hesitation of governments to take appropriate measures to clean up the waters and atmosphere that are too quickly condemning these animals to death. Such is literally the case of the beluga population of the St. Lawrence!

A little too fast, a little too close?

As the number of whalewatchers and consequently, whale lovers, increases, so will the pressure exerted on the elected officials to take the necessary steps in improving the fate of the whales. After all, the more people see whales and marvel at them (and who is not amazed by the sight of a whale?), the better off whales will be.

To end this chapter on a more positive note, it is important to realize that the governments of various countries are becoming more involved in improving the fate of cetaceans. They are now more aware of the threats whales face, and by the same token, of the considerable economic contribution that ecotourism brings. The moratorium on the commercialization of whale by-products and the pressure exerted on whaling countries for the implementation of very strict quotas are certainly proof of that. The creation of protected areas that, on the one hand, ensure cetaceans a relative tranquillity and, on the other hand, enables us to observe them in their natural habitat, in conditions that respect their wellbeing, is now more widespread. Such areas are almost everywhere. Canada has become a leader in this field. It created the Saguenay—St. Lawrence Marine Park, which encompasses a large part of the area of current upwelling near the mouth of the Saguenay River. Its primary objective is to increase the level of protection of this region's marine ecosystems as well as to promote their use for recreational purposes. Moreover, the preservation of an ecosystem does not necessarily imply preventing public access. Not at all. Only by providing carefully and appropriately supervised educational activities, by increasing public awareness, and by understanding that the environment is truly part of our heritage—only then will we ensure ourselves of a better future on earth— this earth we have *borrowed* from our descendants.

PINNIPEDS

Pinnipeds have been hunted extensively since immemorial times. They are more easier prey than cetaceans because they haul out on land and, generally speaking, they are smaller and more easily handled.

Seal bones have been found at several coastal archeological sites, from upper Paleolithic (28,000 years B.P.). Maybe we could find a lot more of them if the level of the seas had not changed so much since the last glaciation. During ice ages, large areas along what is now the coast became accessible when the level of the seas was much lower than today. Those areas must have been inhabited by humans and there must be many archeological sites now under more than 100 m of sea water. Each year fishermen trawl up hundreds of mammoth, woolly rhino, and other animal bones from the North Sea, which was above water during the last ice age and apparently a good place to live. Near Marseilles on the Mediterranean Sea, a scuba diver found the entrance to the Cosquer Cave, 70 m underwater. This cave, which contained very beautiful paintings, must have been accessible from land at the time.

Shaft straightener, bâton de commandement, Montgaudier. Two gray seals are easy to identify.

Prehistoric tomb of a woman and child found at Västerbjer. Arrows point toward a row of seal teeth probably sewn on the edge of a dress or a coat.

A hollowed-out stick engraved with two seals, a fish and two serpents was found in Montgaudier, in the southern part of France. It is around 12,000 years old. These sticks were used to straighten arrows or assegai shafts, and it is quite possible that these engravings represent animals hunted or fished by the owner of the tool. In a grave in Västerbje dating from the Mesolithic a woman and a child were buried. A great number of seal teeth can be seen in a row and must have been sewn onto the edge of a skirt or a coat, clearly showing that seals must have been well hunted in northern Europe.

The seals provide good meat to eat, a strong skin which can be used in garments, lines and skins for kayaks, and oil for many uses. The fur seals have been hunted too, for their rich fur made them even more tempting. The Walrus has been hunted for its ivory, its skin, oil and meat.

Man has never showed any interest in preserving resources in the past and pinnipeds are no exception. Many species were driven to the brink of extinction and some did disappear (Caribbean monk seal, *Monachos tropicalis*, and the Japanese fur seal, *Zalophus japonicus,* are examples).

The near disappearance and the strong comeback of the Northern Elephant seal is very well documented. Heavily been hunted since 1810, it was clear, in 1860, that it would not be able to survive such treatment very long. In spite of the warning signs, it was still much hunted until 1884, when only about 150 animals were killed. The species was thought extinct, but, in 1892, some 7 or 8 were found on Guadalupe Island and, naturally, killed. This time everybody thought they were the last ones. Happily, it was not the case and some must have survived in remote, inaccessible places, since in 1922, the population was estimated of around 150 animals and it was decided to institute legal protection. Nowadays, they have made a comeback over all their former distribution area and their population is estimated at 150,000 heads. It is a very good example of what protection can do to save a species from oblivion, provided that consanguinity is not too big an obstacle.

There was a big Walrus population in the St. Lawrence Gulf on Magdalen Islands. They were hunted to extinction in the XVIth century. Hunters went to the beaches where the Walrus hauled out for the night, wearing dark clothes and carrying long poles. Quietly, they would poke at some walruses with the tip of their pole. The animal treated that way thought it was another one poking it with its tusks to climb on the beach. It would move up on the beach poking some neighbors with its tusks along the way. The neighbors would do the same, and in short order a very small group of hunters could displace tens of Walruses toward the interior, where they could be killed easily because they were too far from the water to escape. In the beginning, they were hunted for their ivory, skin, and blubber. When their ivory lost almost all its value because Elephant ivory from Africa came in great quantities to Europe, they were hunted for their skin and blubber only, the tusks remaining on the killing grounds. The walruses had disappeared by 1764. Around 1800, a merchant offered $0.01 for any large tooth found on the ancient killing grounds. He gathered 2 tons of ivory in that way! That shows the enormous quantity of the animals killed there, and, by the way, the considerable size of the herd. One can still find skeleton parts near the Grande Échourie" (the Big Haul-out) which was one of the favorite haul-out sites, and, from time to time, fishermen pull heads and teeth from this period from the sea.

Walrus skulls found by fishermen on the sea bottom, near Îles de la Madeleine.

The Greenland seal has been also heavily hunted in Canada, near the Magdalen Islands and Newfoundland. This seals, almost exclusively pelagic, migrate south in early spring, to lay down their pups on the ice and to mate. The pup, or "white coat," is covered by a very fine coat of white fur. Each year, people from the Magdalen Islands and Newfoundland organized sometimes very risky hunts on the ice floes, to hunt for the "white coats." Tens of thousands were killed each year, hit on the head by a heavy club, skinned, and then brought back and loaded on ships trying as well as they could to circulate between the ice floes.

Hunting grew constantly in Newfoundland waters from early in the nineteenth century until 1867. During those years, more than 700,000 skins could be gathered each year, and 7,500 tuns of oil (each tun equivalent to 1.125 liters) exported mainly to England.

Sailing boats and hunting ships, used at the beginning, were replaced by steamers, but when they appeared, the value of the pelts and oil was beginning to drop, and in fact the herd had been depleted, though not in danger. After 1980, following pressure from animalist groups, the markets for the pelts became quite nonexistent and hunting came back to an artisanal state. Recently some commercial hunting has started anew. There is an Asian market for the seal penises, and some market for the meat and skins

As is currently the way the human species reacts, the pendulum went completely the other way, and nowadays, pinnipeds have a special status in a lot of countries where it is forbidden to market any part of a seal, even if populations have attained levels where their survival is not an issue anymore, and even have attained levels where their numbers can cause problems. The Greenland seal herd near Canada, reached more than 5 million heads in 2004 and is still growing annually.

If we consider, in a very conservative way, that a seal eats at least 5 kg of fish a day, we can reasonably estimate that more than 25,000 tons of fish is taken each day from the sea by this species only. We can conceive that it can have a really bad effect on the reconstitution of depleted fish stocks (such as cod, for example). Even if their diet does not include many commercial fish, they eat the same prey.

When we see countries like the USA, which forbids all trade in seals products in their country, but has a president who refused to sign the Kyoto accords on the limitation of emission of green house effect gasses, we can have a good idea about the hypocrisy at the source of these regulations. Green house effect gasses will lead to a rising in the atmosphere temperature. If the temperature rises, and prohibits the formation of the ice around Newfoundland what will the harp seals, which need it to have their pups, do? Recently there has been two years with poor ice floe formation near Magdalen Islands. Hundreds of pups drowned because their mothers evidently could not hold themselves. When we compare the effects of a well regulated hunt and those of an eventual disappearance of the ice floes on the pupping grounds, one can only wonder at the pertinence of the choice of the means used to protect efficiently the seal (and marine mammals in general) populations! The attitude of certain persons, like a well known actress for example, who see nature only through the prism of their emotions, without any implication of their reasoning mind, has a real perverse effect, as we will discuss later.

RESEARCH

Not so long ago, biologists (or naturalists as they were called then) depended on strandings to increase their knowledge of whale anatomy, and on the tales of navigators and whalers to learn about their habits. But the development of onshore whaling stations and floating factories allowed biologists to accumulate more information on cetacean anatomy and physiology, their distribution, and their behavior. Studies of this type still take place on board vessels that practice "scientific" whaling. Although this practice usually conceals commercial interests, it does significantly contribute to our understanding of the biology of the still-exploited species, especially the Minke whale, possibly the only species now capable of sustaining a certain pressure from whaling.

Coastal research stations are numerous. Scientists, often assisted by volunteers, have made considerable progress concerning cetacean biology. Some stations are, to varying degrees, even associated with certain branches of the Navy, and in spite of the dubious nature of some experiments, the financial means these stations are provided have allowed us to further increase our knowledge on the biology and physiology of these marine mammals (see the chapters on diving and on the biosonar).

Studying whales in their natural habitat is extremely difficult. They spend very little time at the surface of the water, and those activities we can witness are limited to behavior patterns such as those associated with reproduction and feeding. Also, whales cover immense territories and are difficult to identify, unless obvious scars or color patterns are present. Finally, some species usually shun coastal waters or are very seldom seen, so learning about them is extremely difficult. This problem is accentuated when the whales concerned are few in numbers, very timorous, or easily confused with other species. For instance, Longman's beaked whale (*Mesoplodon pacificus*) is known from only a couple of skulls; one found on a beach near Mackey (in the state of Queensland, Australia) in 1882, which was identified by Longman as a new species in 1926, and the other on the floor of a fertilizer factory in Mogadishu, Somalia, back in 1955. It came from a beach near Danané (Somalia). Since then, possibly two more specimens were seen near the Seychelles, in 1980. Fortunately, many species are coastal or come close to the shores at one time or another during the year.

To study a population of whales, one generally begins by taking its census. Some species, such as the California Grey whales, have to go through a very narrow strait, the Unimak Pass, in the Aleutian Islands. It is easy for an individual willing to face the rigors of the climate to count all the Grey whales that pass before him. This is precisely how we managed to determine this whale's numbers accurately and to take it off the endangered species list. This case is exceptional, however. Taking a census of aquatic species is generally difficult, and cetaceans are no exception to the rule. These difficulties are best illustrated by the case of the Beluga in the St. Lawrence River.

Beluga in the St. Lawrence River. The markings on its side are not necessarily useful in identifying the individual. Superficial markings such as these can disappear with time.

This population separated from the main herd toward the end of the last glacial period and now occupies a narrow, geographically well-delineated, and easily-accessible territory. Despite the numerous efforts and the substantial financial input in inventory activities, we have only an estimate of this whale's numbers. There are "about" 1200 individuals and we "believe" that this population is stable; it "seems" that the number of births compensates for the number of deaths. Imagine now the difficulties that the cetologist faces in trying to access the number of Right whales, Sei whales, or Northern Bottlenose whales!

When conducting inventories, it is very difficult not to count the same individual more than once since it is constantly moving and not much of its body is seen at the surface of the water. It is therefore necessary to find ways of identifying individual whales. Fortunately, for biologists at least, cetaceans fight among themselves leaving visible bite marks on their bodies, and sometimes they collide with various objects such as rocks, boats, etc. Permanent body markings resulting from such escapades can help biologists to identify specific animals. This is especially true when scars are on the dorsal fins, tail, back, or even on the animal's upper side. Photographs are taken and catalogued for reference (photo-identification).

Some species show markings and color patterns that vary from one individual to the next, much as our fingerprints do. Identifying individuals within a population is thus made possible. By taking thousands of photographs, we have learned that some whales return to the same location year after year in order to breed or feed. In Québec, *le Groupe de recherche et éducation sur le milieu marin* (GREMM) has compiled an impressive collection of photographs of Belugas and of Fin whales. Richards Sears has done the same with the Blue whales of the Mingan island area. Ned Lynas' Ores Coastal Study Centre at Grandes-Bergeronnes (North-Shore of Québec), through a photo-identification program, is taking a census of the Minke whale population, truly an achievement as far as this small and very active species is concerned! In Newfoundland, John Lien compiles photographs of Humpback whale flukes. On the West Coast, through similar photo-identification projects, the social structure of Killer whales is now better understood.

Do you recognize them?

One page in the catalog of Minke dorsal fins as assembled by ORES group.

"Zipper," a Fin whale easily identifiable by a scar caused by the propellers of a boat.

Flukes of a Humpback near Percé (Quebec).

The tail of a Humpback whale, near Percé, showing its distinctive color pattern. Note the contrast with picture above.

"Double scoop."

In Texas, the photo-identification of Bottlenose dolphins ensures a more accurate census, as well as better study of their movements and the stability of their social groupings. Photo-identification is becoming a generalized inventory taking method, applicable to any cetacean species that, at one time or another, approaches the shores, tolerates the presence of boats, and stays at the surface long enough for pictures to be taken. Throughout most of the world, organizations dedicated to cetacean research develop similar methods specifically adapted to the species studied.

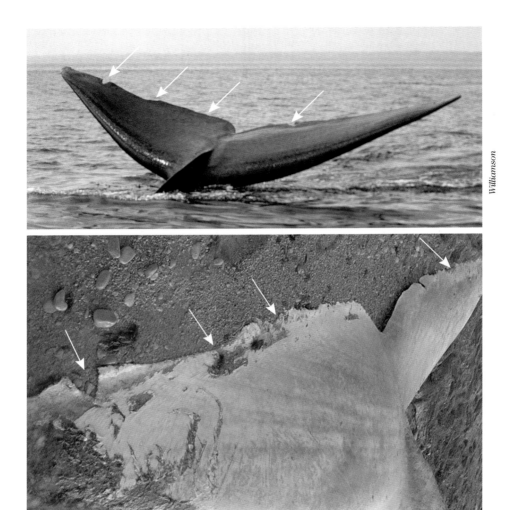

"Pita," a Blue whale that was a thirteen-year summer resident of the St. Lawrence River. Photographs of Pita's flukes were used to positively identify its carcass (July 1993, île d'Anticosti).

Many species travel considerable distances between their breeding and feeding grounds. As a result, they are more difficult to detect in the open sea far from the shore. A rather unusual way to follow Pacific Blue whales is to use the detecting devices of submarines. In fact, the U.S. Navy has a network of buoys that keeps track of submarine movements. As we have mentioned earlier, Blue whales emit extremely powerful sounds that travel great distances. Each animal has, in all likelihood, its own acoustic signature. By triangulation, we can detect and follow these peaceful mammals' movement on maps—an odd way for mammal research to benefit from military equipment! (See the chapter on migrations). Radio-tracking techniques, involving tags that are usually fixed with suction cups on the back to the great whales, are also employed. These transmitters then send, by satellite, important and accurate data concerning dives (frequency, length, depth) and movements.

Blubber, skin cells, and DNA samples (biopsies) are now used to determine sex and to accurately identify individual animals. Cetologists are trying to establish the genetic map of many cetaceans. This is made possible through biopsies performed by the use of specially-designed arrows that take skin and blubber samples or by simply collecting skin fragments that are constantly shed by whales (a little like dandruff). Consequently, the DNA collected from the individuals within a population can be used to determine the consanguinity within the group and eventually improve our actions in the protection of threatened whales. The Norwegians take DNA samples from each Minke whale they legally kill. This is done to control the selling of poached meat (assuming that control is adequate).

As we saw in the previous chapters, much work remains to be done. Theories concerning the feeding habits of many species, the production and reception of sound, osmoregulation, reproduction, distribution, the physiology of diving, the effects of pollution on the immune and reproductive systems, and other topics are far from complete; in addition, funding for essential research is dwindling. Having said that, I am nonetheless impressed by the progress we have made regarding our understanding of whales since I started working on the second edition of this book. The electronic highway gives us instantaneous access to knowledgeable people from all over the world and to an unprecedented wealth of information. How wonderful!

Blue whale, Percé.

Fin whale.

Belugas.

Only one thing bothers me, though—the perpetuation of the myth that aims at making whales distinct from all other animals, and equal if not superior to human beings.

The mythologizing of whales perversely endangers their very survival. For example, rescuing a few stranded whales or two or three specimens caught by the ice are isolated and highly spectacular events, but really only allow governments and organizations devoted to whales to soothe their conscience. These highly publicized actions involving only a few individuals have but an extremely minimal effect on the survival of cetacean populations. Furthermore, they require considerable energy (most often, that of volunteers) which is diverted from the real problem at hand, cetacean survival. It is the pollution in the oceans that kills and will kill more whales than will the few hunters that practice whaling today. The overexploitation of marine resources through intensive fishing methods, like the use of drift nets that indiscriminately capture and kill animals, certainly have more disastrous consequences on the survival of whales. However, in the current economic situation, such subjects cannot be dealt with directly if one hopes to get re-elected. Consequently, a politician prefers to perform a spectacular (not to mention symbolic) gesture, which leads us to believe that he really cares about the issue of saving whales, even though such a gesture has had an extremely limited impact. Hervé Kempf's *La baleine qui cachait la forêt: Enquête sur les pièges de l'écologie'* (*The Whale That Hid the Forest: A Study of the Pitfalls of Ecology*) clearly illustrates this situation.

Such mythologizing shifts the focus of governments and fund-raisers toward goals that, thanks to the efforts of a few pseudoscientists, are emotionally important to the general public and politically profitable at the same time. One has only to surf the Internet to see how many groups are obsessed by finding ways to communicate intelligently with cetaceans, and how much money they are investing in what seems to me a pure utopia—so much energy and money wasted on chimeras, while we privatize organizations devoted to environmental protection and continue to allow agriculture practices to pollute surface and ground waters.

Man's cupidity has already brought about the overexploitation of whales; let us not give him the opportunity to mythologize them and further destroy their biotope and his.

HISTORICAL PHOTOGRAPHY

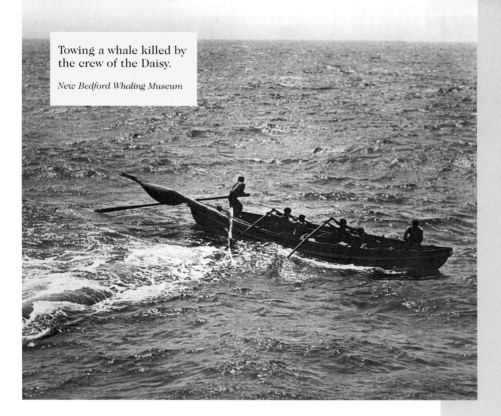

Towing a whale killed by the crew of the Daisy.

New Bedford Whaling Museum

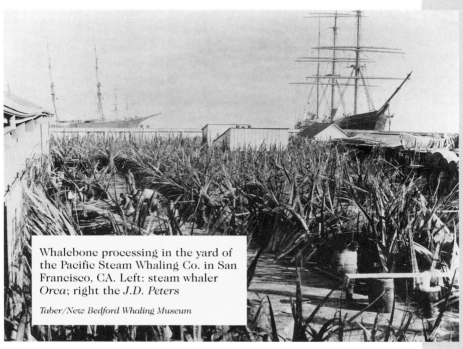

Whalebone processing in the yard of the Pacific Steam Whaling Co. in San Francisco, CA. Left: steam whaler *Orca*; right the *J.D. Peters*

Taber/New Bedford Whaling Museum

The bite of a Sperm whale—water color by unknown whaleman.

New Bedford Whaling Museum

1885

New Year.

Thursday. Bent new main top sail

January 9th Friday.
Gamed Barks Sagoda and Lancer.
Lat 38°04' S.

January 11th Sunday.
Gamed Bark John Carver.
Lat 38°08' S. Long. 16°40' W.

January 20th Tuesday.
Dick — (Foremast hand) raised a
school of sperm whales. we lowered
the three Larboard boats. and the
mate struck one and saved h[im]
Lat 36°55' S Long. 17°30' W

January 21st Wensday.
Blowing a gale. Saw a large sperm w[hale]

January 25th Sunday.
Pete Snyder (Boatsteerer) raised a school
of sperm whales going to the windward.
Lowered the 1st 2d & 4th mates. and the 4th mate stru[ck]

1885

January. 29th Thursday.
  Stowed down 105 bbls sperm oil.
  Land in sight.   Lat 36° 56' S.

February. 8th Sunday.
  Gamed Ship Horatio.
        Lat 36° 20' S.

February. 24th Tuesday.
  Raised a school of Black Fish and we
  lowered the three Larbord boats and
  got three, which made about 3 bbls.

March 1st Sunday.
  Gamed Bark Lancer.
        Lat 36° 42' S.

March 2nd Monday.
  Spoke Bark Legal Tender. of
    San Francisco.   Lat 36° 08' S.

March 5th Thursday.
  Sighted the Island of Juan Fernandez.

A page from the log of the
John P. West, 1885.

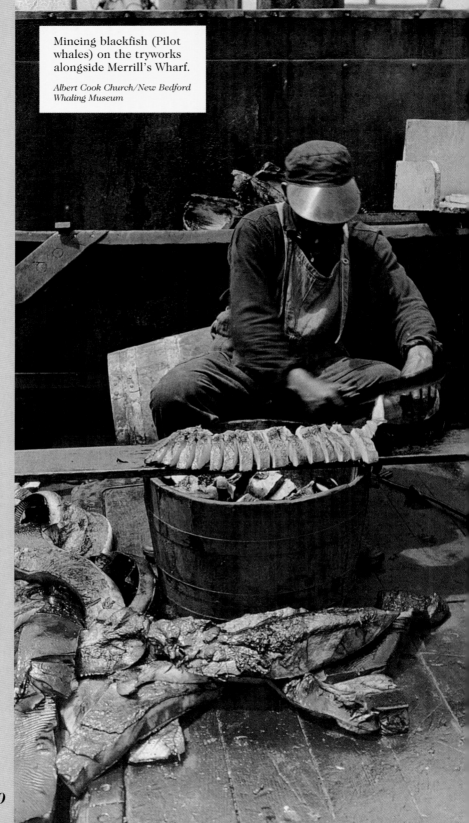

Mincing blackfish (Pilot whales) on the tryworks alongside Merrill's Wharf.

Albert Cook Church/New Bedford Whaling Museum

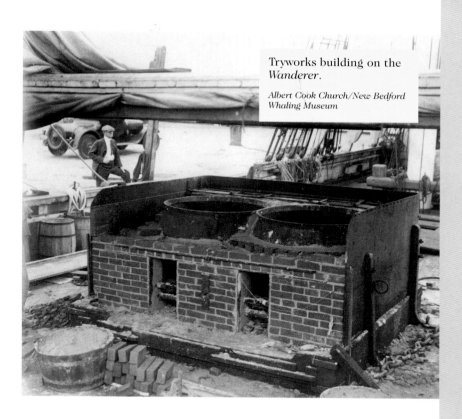

Tryworks building on the *Wanderer*.

Albert Cook Church/New Bedford Whaling Museum

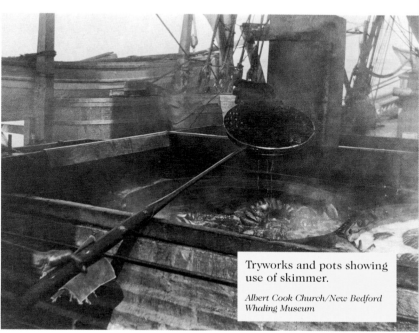

Tryworks and pots showing use of skimmer.

Albert Cook Church/New Bedford Whaling Museum

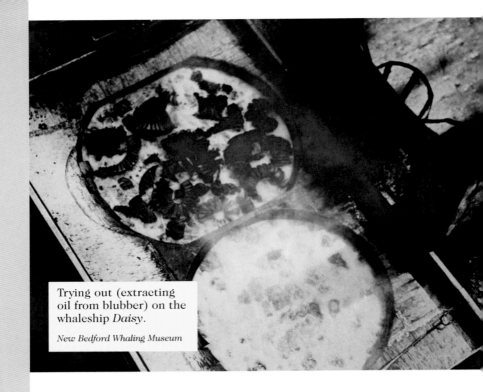

Trying out (extracting oil from blubber) on the whaleship *Daisy*.

New Bedford Whaling Museum

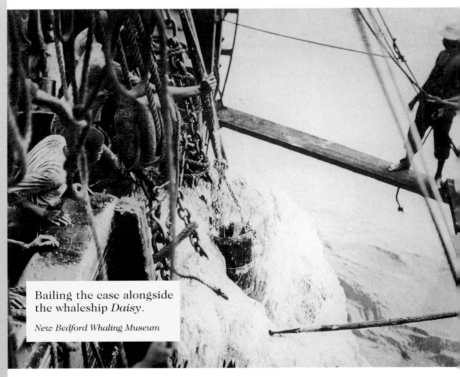

Bailing the case alongside the whaleship *Daisy*.

New Bedford Whaling Museum

Boarding the carcass of a small Sperm whale.

Clifford W. Ashley/New Bedford Whaling Museum

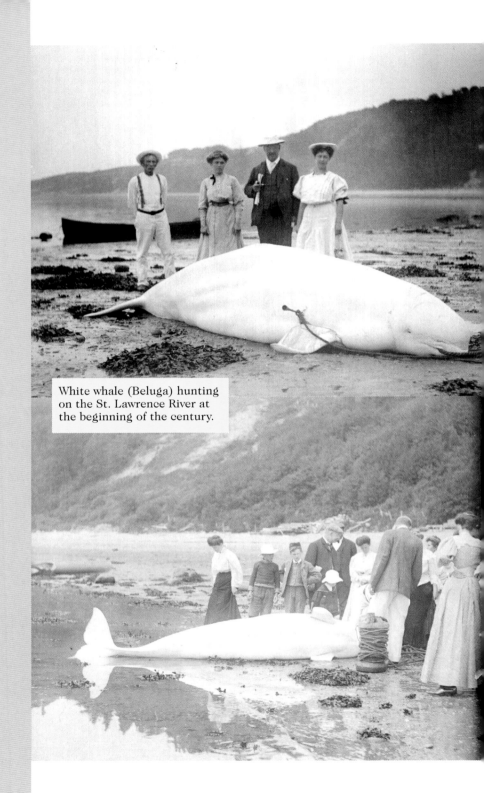

White whale (Beluga) hunting on the St. Lawrence River at the beginning of the century.

PALEONTOLOGY

The sun has reached its zenith; the estuarine water is calm and warm. On the spits of land, the large trees spread their leaves that rustle in the light breeze. Perched on these trees are multicolored birds. On the ground, insects, reptiles, and small mammals chase or flee one another. Without warning, a gigantic *titanothere* appears shaking its muzzle peculiarly adorned with flattened bony protuberances. Its minuscule eyes blink in the intense light.

Suddenly, a powerful force disturbs the calm waters; a long and narrow head surfaces, nostrils open… a mighty blow is heard. Undulating slowly, the animal approaches a sandbank and, leaning on its atrophied forelimbs, partially hoists itself up. A yawn stretches its huge mouth exposing well-differentiated, fang-like teeth.

The long serpentine body has a smooth and swarthy skin: it is definitely not a reptile. The extremity of the body that remains submerged is transversely flattened. Two hardly visible stumps suggest a quadruped ancestor.

Its head resting on the sand, *Basilosaurus*, the "king lizard" falls asleep digesting its latest victim.

We are on the shores of the Tethys Sea (Eocene), forty million years ago, and we have just met our first archaeocete.

CETACEANS PHYLOGENY

Aristotle classed them with the fishes and they remained there for a longtime, maybe with some hypocrisy, so people were able to eat their meat and fat even during the days of abstinence, which were very numerous at a certain time. The fat from cetaceans was even called "Lent fat" during Middle Ages. Once brought back with the mammals, the big problem was where to place them correctly in the genealogical tree.

Cetaceans are so specialized, so well adapted to their way of life that it is impossible to imagine the appearance of their ancestors, so much so that creationists use them regularly as proof of the inanity of the very concept of evolution! All mammals which have lived before and during the first ten millions years (MY) of tertiary period (paleocene) were terrestrial tetrapods. Ancestors of the cetaceans are therefore surely land animals, but it is impossible to determine which when you look at the profoundly modified body of their present descendants.

PROBLEMS WITH FOSSILS

The cetacean paleontology is not well known, because there are not many early fossils. This can be the result of two factors: the environment in which they (and their presumed ancestors) were living and the fact that cetacean evolution was relatively rapid and took place in a not very large geographical zone.

Environment: Cetaceans live and die in the sea. Before sinking to the bottom they float, bloated, for some time on the surface. During this time decomposition may cause the loss of certain parts of their skeleton. When they sink and drop to the bottom where sediments to cover them, they usually are not complete, and can be destroyed by bottom feeding animals. Those which strand rapidly are very often destroyed or have their skeleton dispersed by tide and waves. The conditions necessary for the conservation of a complete and well preserved skeleton are really exceptional.

Geographical area of evolution: Their evolution took place around the Téthys Sea (Mesogea). This sea was situated in an area occupied by the Mediterranean Sea (which may be all that remains of it) stretching to Asia; at the time, India was not attached to the continent. This ancient sea has played an important role as an evolution and dispersion center for a great number of species. If cetaceans evolved in small scale areas, like estuaries or shallow water around archipelagos, it is understandable that we do not find many fossils because their resting places are not accessible or have simply not yet be found.

Politics too can make things more difficult when one wants to find fossils, as surprising as it can seem. Among the richer cetaceans fossil yielding areas are the sediments which were at a time at the bottom of the Tethys Sea, and which have been brought to the surface as Himalayas were formed when India collided with Asia. It is mainly in Pakistan that the geological stratas that are most likely to contain archeocetes fossils are the easiest to explore. The political situation of the region makes access to the sites impossible at times, so the actual digs can be made only in safe periods and that makes the discovery of fossils the more difficult!

Cetaceans evolution took place during the cenozoic period of earth history, beginning 65 MY ago, with the extinction of the dinosaurs (KT extinction so called, because it happened between Cretaceous and Tertiary periods. K from greek *kreta* meaning chalk) and in which mammals experienced a spectacular rise. It began at the end of Lower Eocene (52 MY) and, around Middle Myocene (between 10 and 13 MY) most actual families were present in the oceans, though that does not mean their evolution was finished.

The building of a hypothesis of the history of different species (*Phylogeny*) is based on a science called *systematics* that also includes taxonomy, which describes, identifies, names and classifies living or fossil species. Systematics allows us to determine if characteristics present in two or more species have been inherited from a common ancestor or are simply the effect of convergence during their evolution.

Most systematicians, nowadays, use a method of classification founded by Hennig, based on the research of parental relationships using the state of shared derived characters (synapomorphies) called *cladistic*. If two organisms share the same derived characters, it is more likely that they have inherited it from a common ancestor instead of having evolved completely independently the same character. As far as classification of living beings is concerned, a *character* is a hereditary attribute of an organism, which can be organic, pysiologic, behavioral or molecular. The *character states* are the different forms of a given character, an ear adapted for hearing in air or water, specialized teeth or all similar teeth for example. One can evaluate the evolution of a character state when the passage from a *primitive (plesiomorphic)* character state to a *derived (apopmorphic)* character state in more recent species can be observed. To determine the state of a characteristic among the members of the studied group, one will compare them with the characteristics present among members of a related group (group B). One can infer that if the same characteristic is found in group B, which is thought to be related to the group under study, and in group A, this characteristic can be considered as ancestral to group A. A shared ancestral character cannot be used to trace parentage between the member of the group: hair cannot be used to classify mammals since all have hair. But those same hairs as a derived characteristic can be used to classify the mammals among the vertebrates. When the character state is established, when it clear which are the ancestral character and which are the derived, it is possible to determine the direction of evolution.

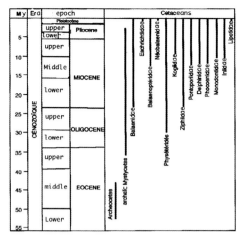

Time scale of living and extinct cetaceans.

If a group contains the common presumed ancestor and all its descendants, it will be called *monophyletic*. On the other hand, a group which has several ancestors will be called *polyphyletic*. A *paraphyletic* group will contain the common ancestor but only part of its descendants.

We are sure, now, that the cetaceans form a monophyletic group, even if there are very big differences between Odontocetes and Mysticetes living today. It means that they all share a common ancestor that they share with no other mammals. We are sure that this ancestor has existed, even if we will never know which it was.

247

CETACEANS CHARACTERISTICS

It is possible to characterize cetaceans by using morphologic characteristics. For example, the presence of a sigmoïd apophyse in front of the external auditory meatus, tympanic bulla and the incus (ossicle in the middle ear) made of dense bone only, without cancellous bone (*pachyosteosclerotic*), and rotation of the middle ear bones are all shared derived characters defining the group, even if they are in a rudimentary state in Pakicetus. This method of characterizing a group can evolve with the discovery of new specimens.

They can be characterized by using the common ancestor approach, which is more constant because there always will be a common ancestor for cetaceans.

To find this ancestor, two methods can be used:
— **The use of morphological characters,** like the way the ear is built, or the dentition, for example, in modern and fossil species.
— **The use of molecular characteristics**, like DNA or the presence of certain antibodies identified by their reaction in presence of certain antigens in living species only.

There has been lots of controversy about the superiority of one method over the other, but more and more, now, they are used in a complementary way.

In 1945, Simpson, an eminent paleontologist, trying to classify cetaceans, thought that their evolution from land forms was so ancient that we would never be able to find their ancestors. It is true that at that time, the only available fossils were those of Archeocetes (B*asilosauridae, Dorudontidae*) which are very much specialized and very well adapted to marine life.

At that time, too, a new approach to comparing living organisms appeared. By using reactions between antibodies and antigens, the degree of parental links was determined. The more reactions to the same antigens in two groups the more related they were.

As early as 1960 this method was applied to different groups of mammals, and it became evident that the group most closely related to cetaceans was the one including the even-toed (2 or 4 digits) ungulates (artiodactyls). In other words, a cow or an hippopotamus were more

Recreation of the skull of *Pakicetus inachus*

Pakicetus inachus, first specimen found.

related to whales than to seals, for example. The question was to know if whales were artiodactyls or if the cetaceans and the artiodactyls had the same ancestor.

Some digging made in Pakistan has led to the discovery of part of a skull and mandible and some teeth from an animal which was called *Pakicetus inachus*. It lived during the Eocene, some 48 MY ago. The way its ear was built, particularly the middle ear, foreshadowed the ear of archeocetes, even if it was not adapted to hearing in the water. The teeth, too, linked it to more modern archeocetes and to a group of carnivorous ungulates, the *Mesonychids* (condylarthras). As no other part of the skeleton was known, and because of its teeth, Mesonyx, a mesonychid, was considered as the cetacean's ancestor.

Then more and more of those anatomically intermediate forms were discovered which were without doubt ancestors of the Archeocetes, but which were not completely tied to an aquatic medium.

With a linear view of evolution all those intermediate forms will be interpreted as "links" in a chain made of ancestors-descendants.

Mesonyx obtusidens.

But, with a view of evolution as a tree, which is the more realistic approach, it will become clear that:

1) Several anatomical intermediaries are not evolutionary intermediaries, but are on collateral branches of the tree.

2) We have no sure method (or even a robustly scientific one) to distinguish between "fathers" and "uncles" among the fossils.

Pakicetus attocki.

It is again in Pakistan, that J.G.M. Thewissen uncovered *Ichtyolestes pinfoldi* and *Pakicetus attocki*, two pakicetids from the lower Eocene, and *Rhodocetus balochistanensis* and *Artiocetus clavis* were discovered by Philip D. Gingerishin and dated to 47.5 MY old. The skeletons, this time, were almost complete, the limbs in particular.

Rodhocetus kasranii.

It was thought that the Pakicetus was as amphibious as are tapirs nowadays, which is not very much. The finding of two hip bones (innominates) have shown that their ilium was a lot shorter than their ischium. The more an animal leads an aquatic life, the shorter is the ilium by comparison with the ischium. Maybe Pakicetids were more aquatic than was previously supposed, at any rate more aquatic than tapirs. (Gingerish, 2003)

If molecular techniques had related cetaceans to artiodactyls without any question, morphologic evidence was less convincing. One of the more distinctive of the artiodactyl characteristics is the shape of one of the small bones in their ankle that people playing knuckle bones know very well: the *astragalus* This bone, which unites the foot to the tibia, in artiodactyls has a pulley (trochlea) at each end, with two deep grooves that keep the movements of the foot in one plane only.

Astragalus from an artyodactyl. The arrows point the two trochlea.

It is viewed as an adaptation for running. The astragalus of more recent archeocetes was too much modified to be used in this way, but those of the pakicetids, Artiocetus and Rhodocetus, did not leave room for any doubt. Cetaceans were really artiodactyls. They just joined Ambulocetus natans as potential cetacean ancestors. It had been found by Thewissen in 1994 and its almost complete skeleton had allowed us to have a good idea of the appearance of an amphibious cetacean ancestor. Its astragalus, however, in a bad conservation state, did not allow for linking it morphologically to the artiodactyls without some doubt.

Astragalus from an artiodactyl ungulate, a perissodactyl ungulate and a carnivore.

By the way, Mesonyx has lost its position as a cetacean ancestor, since its astragalus, cannot be related to the artiodactyls. To consider it as an ancestor, it would be necessary to admit that its astragalus had reverted the tendency leading to the shape of artiodactyls during its evolution. What had been interpreted as ancestral characters (ear structure, teeth) were perhaps only the result of convergent evolution. More fossils will have to be found to close this question definitively.

The early artiodactyls form, which could be the common ancestor both to artiodactyls and cetaceans, remains to be found. Diacodexis (lower Eocene) is the oldest artiodactyl known today based on the shape of its astragalus. It cannot however be directly linked to cetaceans.

The *diacodexis* is considered as the oldest known atiodactyl.

There exists among artiodactyls a group known as the Anthracotheres, and among them an animal called *Elomeryx armatus* (brachyrhinchus), whose teeth and the fact that its skeletons are usually found in fluvial deposits suggest that they were living a semi-aquatic life, a little like the hippopotamus today (Kron and Manning, 1998).

Elomeryx is too recent to be an ancestor to cetaceans. It lived in lower Oligocene (30 MY). It is quite possible, however, that a similar but more ancient form of anthracoteres was the cetacean ancestor (Gingerich et al., 2001). This form is still unknown, and here again, more fossils are needed to clarify this question. We could see among those forms which one might be the common ancestor of the cetaceans and the hippopotamus, which for the time being seems to be the land mammal nearest to the cetaceans.

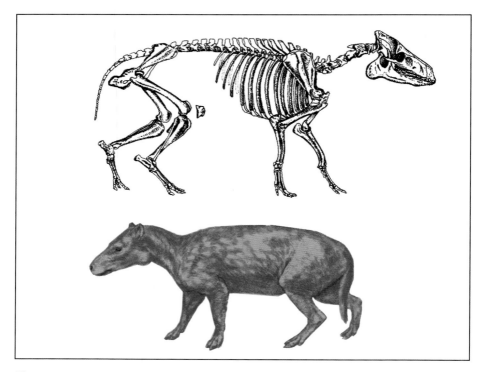

Elomeryx armatus.

EARLY ARCHEOCETES

What did those early archeocetes look like?

Pakicetids were tetrapods 1.80 m long at most. Their limbs were well adapted to running and jumping, their astragalus was typically artiodactyl, and their feet were paraxonic, which means that their weight is supported mainly by the fingers 2 and 4 (or that the axis of symmetry passes between those two fingers). In perissodactyls, the feet have odd numbered fingers and the axis of symmetry passes by finger number 3. The feet are called mesaxonic. The morphology of the last phalange of their fingers shows that they had hooves instead of claws.

Last phalanges for hoof or claws. **Top row,** ungulates phalanges: donkey (Asinus sp.); moose (Alces alces); and sheep (Ovis aries). **Bottom row,** phalanges bearing claws: bear (Ursus americanus); seal (Halichoerus grypus); and lion (Panthera leo).

The anatomy of the ear does not reflect a great adaptation to an aquatic life, even if the ear bones of the middle ear have begun their rotation, but shows that they may have used their mandible to collect the vibrations of the ground as do some modern reptiles, crocodylians for example.

ADAPTATIONS TO AN AQUATIC LIFE

One can wonder why land animals have become gradually more and more adapted to an aquatic life. Several reasons can be called upon. When the dinosaurs and the huge marine reptiles disappeared, they left vacant many ecological niches on land and in the oceans. This situation led to an explosion of mammal populations and a rapid diversification of existing mammal forms. More and more heavy competition for food on land probably pushed some species living at the edge of the sea to forage in shallow water at first, then moving deeper and deeper. For any species to be able to lead a more aquatic life, it was imperative that natural selection worked on existing structures which could be bettered in this direction. In such an environment, it is evident that it was necessary to have the ability to swim and catch aquatic prey. The ancestral forms must have endured pressures that would promote a greater ability to swim.

The study of the cetaceans labyrinthic system of the internal ear, in which the semi-circular canals (used to maintain equilibrium), shows that they have a shorter radius than what can be found in land mammals.

Beluga's labyrinth partially dissected. One of the three semi-circular canals is visible; very small compared to the animal's size. (3 to 4 m).

Greenland seal labyrinth. The semi-circular canals are a lot bigger. It could be due to the greater mobility of the head.

    The effect of this is to reduce the sensitivity of the system by quite a lot. It would correspond to an adaptation for more acrobatic movements in the water (the effects of gravity are less perceptible in water) and compensate for their very short neck which does not allow them to keep their head stable during these movements (in fact a very mobile head compensates for a greater sensitivity of the inner ear, as we can see when we look at the jerky movements of the head of a ballet dancer or a figure skater when they spin quickly.) Those modifications appeared very early in the cetacean ancestors, maybe only 5 MY after they had diverged from the common ancestor. This specialization was, perhaps, the point of no return in cetacean evolution; natural selection could not but favor more and more aquatic forms, their new labyrinthic system being an obstacle to their well being on land. (F. Spoor et al., 2002) This is why new forms, more and more adapted to an aquatic environment, make their appearance in a relatively short time

The Ambulocetids were more than 4 m long and must have walked on the ground a little, like fur seals do today. When in the water, they must have used their feet, which were large and webbed, as paddles, maybe as Desmans do (big semi-aquatic shrews or muskrats). Their ear was not very specialized toward underwater hearing, but began to be isolated from the skull. It is speculated that they must have also laid in wait, half in the water, the head flat on the ground, monitoring the vibrations made by an approaching prey. When it was near enough, they must have jumped on it, grabbing it with powerful jaws and pulling it into the water to drown it.

The Protocetids, among which we find *Rhodocetus*, looked like *Ambulocetus natans*, but some had bigger feet and hands that made them better swimmers. They used their four limbs to swim and their tail was used to dampen the oscillations caused by the alternate movement of their well developed feet while swimming.

DEVELOPMENT OF THE TAIL AS A PROPELLING TOOL

As early as 1883, it was suggested by Flowers that the whale's ancestor must have had a long tail. This hypothesis does not fit well with the fact that cetaceans are descendants of an artiodactyl, which, in general, have a short tail. How was it possible that the tail became the main and finally the only propeller in whales?

If we consider that the artiodactyl ancestral form is similar to Elomeryx, the first stage toward water locomotion would be the development (as we see it in Protocetids fossils) of big webbed feet paddling in alternance. This way of swimming would promote the development of forms with long tails since it is used to dampen the oscillating movements due to the of alternate movements of the feet, as we have said earlier.

REMINGTONOCETIDS

The Remingtonocetids were smaller. Their skull, well known, is quite different from the skull of the other groups. By their teeth, we know that they must have hunted fast swimming prey, but by laying in ambush because what we know of the rest of their skeleton does not seem to have made them fast swimmers. They did not exist for very long and apparently left no descendants.

Kutchicetus Remingtonocétidés

KUTCHICETUS, REMINTONOCETIDAE

It seems that all these primitive four-limbed archeocetes had disappeared before the end of the Eocene, around 40 MY ago. Many of their characteristics show clearly their parentage by artiodactyls. During a good part of Eocene, there were multiple forms more or less adapted to an aquatic life, but without any being completely tied to the water. We will have to wait for the Basilosaurids to have completely aquatic forms.

BASILOSAURIDS

The Basilosaurids include the Basilosaurinae and the Dorudontinae.

Basilosaurinae have been known since the 19th century. The first vertebrae of this animal were thought by Harlan, in 1832, to be those of a reptile. He gave it the name of Basilosaurus (King serpent) because they were very huge. However, Owen recognized it as a mammal and proposed for it the name Zeuglodon cetoides as early as 1835. Other parts of its skeleton, found in 1845, lead Wood to put it with the cetaceans.

A little later, a German collector, having heard about the beds containing Basilosaurus, came to Alabama to do some digging. He unearthed many bones from many individuals, and finally put them all together to build a skeleton more than 34 meters long. He exhibited his work in several European cities, calling it "the great sea serpent" He was so good at presenting the specimen that the Prussian king bought it for the King Anatomical Museum in Berlin (which gave Koch a good annuity!). There, in spite of the denigrations of the former owner, the composite skeleton, with all the errors made in the reconstruction of the specimen, was unveiled by the museum director, J. Müller.

Since then several specimens have been found in United States and in Africa, mainly in Egypt in the beds of "Wadi Hitan," "Zeuglodon Valley" also called "the valley of the whales." Hundreds of skeletons of basilosaurs lie there, complete enough to have their hip bones and hind limbs preserved. Those are very small. They may be only vestigial remnants without any use, but some details in the way the knee is made, could have made of those limbs tools to hold the partner during copulation.

Zeuglodon skeleton, Smithsonian Institution.

Archaeocete skeleton, Wadi Hitan, Egypt.

Francis Duranton

Basilosaur.

Some Basilosaurs were very long and snake like, growing to more than 20 m. The aperture of the nostrils was still halfway up the snout, half way to the position they occupy in modern cetaceans. Their teeth were not all alike, but still differentiated into pointed incisors, in two parallel rows at the tip of the rostrum, and multituberculate cutting premolars and molars toward the rear of the mouth.

Archaeocete skull, Wadi Hitan, Egypt.

Francis Duranton

Their vertebrae, especially the lumbar and caudal, are extremely elongated. It is a very specialized characteristic, which eliminates them as direct ancestors of the modern whales. One of their caudal vertebrae has a shape identical to the one situated in the tail of modern cetaceans, where the flukes begin. The "ball vertebra" is situated between the preceding vertebrae whose greater axis is vertical, and the last one (those in the flukes) whose greater axis is horizontal. This vertebra,

Caudal vertebrae of the Beluga. The arrow points the "ball vertebra."

more spherical than the others, gives a greater mobility to the flukes which are the only propeller in cetaceans. One can conclude that the basilosaurs moved through the water, as modern whales do, by using their flukes.

The Dorudontinae were relatively small animals the biggest found not exceeding 6 m. The best known belong to the species Dorudon and Zygorhiza. As was the case with the basilosaurs, their nostrils are still midway on their rostrum and they still had teeth recognizable as incisors (pointed for grabbing), canines and premolars and molars (made for cutting). They also have in their vertebral column the "ball vertebra," the spherical vertebra that is the proof that they used the horizontal flukes of their tail to swim.

Zygorhiza.

Dorudon.

Their vertebrae do not show the lengthening typical to the basilosaurinae, so it is maybe in this group that the common ancestor of Odontocetes and Mysticetes can be found. However some details of their dentition raise doubts about that. Here again, more research and fossils are needed to clarify this.

The Archeocetes made their appearance about 50 MY ago, near the end of the lower Eocene, and by the end of the Eocene period, about 33 MY ago, they had all disappeared. In this group we find all the presumed cetacean ancestors that have kept a certain mobility at the level of the elbow joint and do not show any telescoping of the skull.

This peculiar disposition of the bones of the head, which is a characteristic of modern cetaceans, appeared with the archaic Odontocetes and Mysticetes. It is the result of the lengthening of premaxillaries and maxillaries which have followed the progressive move of the nostrils toward the top of the head. This telescoping of the bones of the head did not occur the same way in Odontocetes and in Mysticetes. Because of that, it is possible to distinguish the fossils of the two types of whales.

ODONTOCETES:

A) The maxillary bones completely cover the frontal bone (the part of the frontal which makes the upper margin of the orbit called supraorbital process) In that way the frontal bone is almost completely occluded when the skull is seen from above.
B) When the skull is seen from above it is asymmetrical. There is a certain level of torsion of the bony nares toward the left.
C) There is a foramen on the premaxillary bone.
D) There is an anteorbital notch at the base of the rostrum, a passage for facial nerves.

MYSTICETES:

A) Maxillaries do not cover the supra orbital process of the frontal bones but send a shelf under the orbit of the eye and a lateral anteorbital process.
B) The occipital bone makes a triangular and oblique shield toward the front.

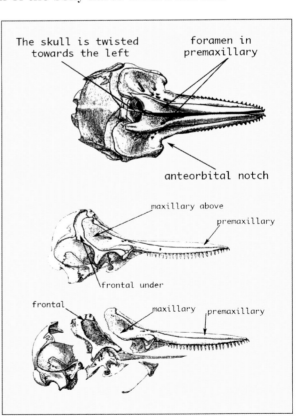

C) The junction of the two mandibles is never welded, but articulated in a way that allows for the rotation of the mandibles in a horizontal plane when the mouth opens (rorquals).

The division of cetaceans in Mysticetes and Odontocetes happened about 30 MY ago. The oldest physeterids fossils known are at least 21 MY old. Some works implying DNA give a date of 25 MY.

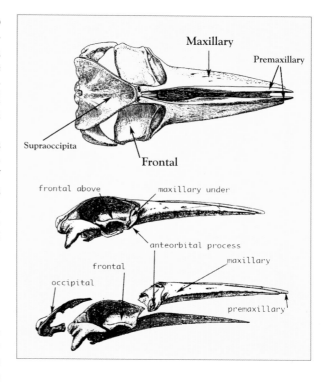

ARCHAIC ODONTOCETES:

Five families are known. The most ancient (*Agorophoriidae*) are known to have existed since the Ligocene (28-24 MY). They are the first to show some degree of telescoping in the skull. Their nostrils are still anterior to the orbit of the eye.

The *Squalodontidae*, known from upper Oligocene to upper Miocene, have a skull almost modern in shape; the telescoping of the bones of the head is almost completed. Their dentition is still heterodont, that is they still have incisors for grabbing at the tip of the rostrum and cutting molars, triangular in shape, a little like shark teeth (hence their name) further in the mouth. There are also the *Squalodelphidae*, the *Dalpiazinidae*, and the *Wapatiidae*.

Squalodon culvertensis (16-10 My). Smithsonian Institution.

Among the modern Odontocetes families, the most ancient must be the Physeteridae with the Sperm whale being the only member left. It could be as old as middle Miocene (16 MY) The other families are known since upper Miocene (11 MY).

Eurhinodelphis sp., a primitive dolphin (16-10 My) Smithsonian Institution.

MYSTICETES

There is no doubt about the fact that Mysticetes have a common origin with Odontocetes and that they share with Odontocetes an ancestor that they do not share with any other mammals.

It is from a toothed species that Mysticetes evolved. The appearance, in the oceans, of great quantities of crustaceans like the euphausids perhaps was the triggering factor that determined the selection of forms better adapted to the en masse capture of prey.

ARCHAIC MYSTICETES:

Aetiocetidae, Mammalodontidae, Llanocetidae. Kekenodontdae: they are considered as Mysticetes even if they all have teeth, because the shape of the maxillaries and the way the telescoping of the bones of the head is done do not leave room for another opinion. They are not the ancestors of today's Mysticetes (too much specialized, too late in time). They are known since upper Oligocene (21 MY).

Dr. L.G. Barnes holding the original fossil of *Aetiocetus cotylvaleus*.

Aetiocetus cotylalveus.

Cetotheridae: in this group are several archaic Mysticetes, without teeth, and probably with baleen plates. It is a heteroclit assembly of specimens put together because they do not have the characteristics of today's baleen whales. Their maxillary is flat and has the grooves and foramen typical of Mysticetes associated with the circulatory system needed to make baleen grow. Some fossils were even associated with the imprinting of baleen plates. Therefore, they have been considered as ancestors of today's baleen whales. Their fossils date as far back as the upper Oligocene to upper Pliocene. (21-6 MY)

The head of a primitive Mysticete. Smithsonian Institution.

Parietobalaena, a primitive Mysticete (15My) American Museum of Natural History, New York.

MODERN MYSTICETES:

Balaenopteridae (rorquals) fossils date as early as the upper Miocene (12-10 MY)

Balaenidae (Right whales) date to the lower Miocene (23 MY)

Eschrichtiidae (Grey whale) the oldest known fossil is 100,000 years old

Neoblaenidae (Pigmy Right whale): there is only one known fossil which could be related to this species, but it was impossible to date it.

There is lots and lots of research to do if we want to completely untangle cetaceans phylogeny. This chapter is only a small summary of what is known, today, of the history of their evolution.

The discovery of new fossils, will give us a chance to solve still unknown parts of it, as the nature of the artiodactyl ancestor, the group from which modern cetaceans were born, Physeteridae position, the group from which Mysticetes arose, the origin of baleen plates, the degree of kinship with the hippopotamus, ...the list of the topics that need clarification is very long and will give lots of paleontologists lots of work for many years... What is encouraging and a good omen for the future is the progress made in recent years with the discovery of many new fossils. It is wonderful to see that the more the digging techniques progress, the better we know where to dig, the faster we make headway in the knowledge of the history of the cetaceans, which is also part of ours...

Wadi Hitan, Egypt and its archaeocetes skeletons.
Francis Duranton

PINNIPEDS

We do not know much about pinniped phylogeny. They appear as fossils relatively late in the fossil scene. The first known fossils which are undoubtedly related to pinnipeds are found in Oregon's upper Eocene (27-25 MY) fossil beds. Modern pinniped lineages diverged soon after that with the first phocids fossils appearing in sediments around the north Atlantic.

Some primitive pinnipeds. American Museum of Natural History, New York.

There is no question today that they descended from a group of arctoïds carnivores. This group included *Ursidae, Procyonidae* (raccoon) and *Mustelidae* (weasels).

The first known specimens that can be related to the Otariidae are distributed among five species of the genus *Enaliarctos*. Their dentition is still heterodont, with large shearing molars. Their teeth and some characteristics of their skull link them (as shared derived characteristics) to primitive ursids, *Amphicynodonts*. Certain species of *Enaliarctos* show, however, a tendency toward the reduction of the number of teeth and with a more pointed shape (homodonty).

Enaliarctos measli is one of the best known of this type, since we have a virtually complete skeleton. It was approximately the same size as the Harbour seal (1.2-1.5 m), and must have weighed around 80 kg. Its limbs were transformed in flippers to swim, but the structure of its hind limbs shows that it was well able to walk on land, where it must have spent more time than today's pinnipeds.

Enaliarctos measli skeleton.

Another fossil nearer to today's pinnipeds appeared, in Oregon also, in the early Miocene (19-16 MY). *Pteronarctos goedertae* had a very small lacrimal and the way its maxillary participate to the orbital wall shows a clear tendency to what will become the rule in pinnipeds (no lacrimal, maxillary part of the orbit wall).

Otariids which appeared last on the fossils scene, are usually divided in two subfamilies: the *Otariinae* (sea lions, 5 genera) and the *Arctocephalinae* (fur seals, 2 genera).

For many years, there were lots of discussions about the monophyly or the diphyly of pinnipeds. Some scientists put together Walrus and otariids as having an ursid ancestor, basing their hypothesis on morphological characters. Seals were thought to have a mustelid ancestry.

Recent research has shown that they all descend from an ursid ancestor, but this opinion is not unanimously accepted. While there is general agreement around a remote arctoid ancestor, though it is not yet known, the possibility of a diphyletic origin is still considered as plausible (Barnes, personal communication).

The Walrus poses certain problems of classification owing to a mosaic of characteristics found in otariids (like the possibility of putting its feet under its body) and in phocids (thick blubber, large ear bones, use of the feet for propulsion in the water, no ear flaps). Nowadays, it is placed with the phocids, but more research will be needed to definitively clarify its taxonomic position. Its astragalus is very similar to the astragalus of otariids and would link it to them (Barnes, personal communication). The primitive Walrus date from as far as the middle Miocene (16-9 MY) They have been found in the northeastern part of the north Pacific. They did not have the big tusks so characteristic of their descendants, which appeared relatively recently. Modern Walrus appear in sediments from lower Pliocene (4 MY) in Belgium. They returned to the Pacific 600,000 years ago.

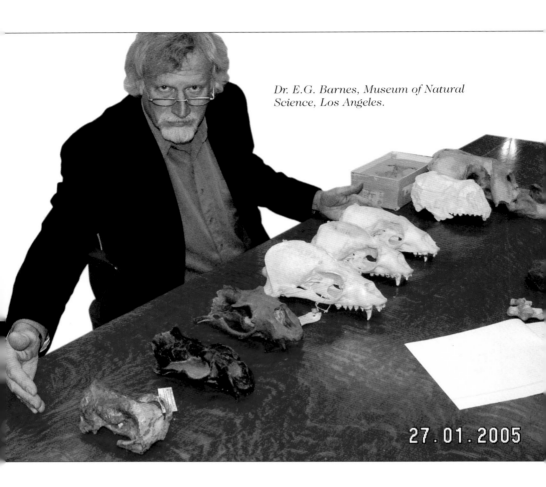

Dr. E.G. Barnes, Museum of Natural Science, Los Angeles.

The phocids, which are classified in two groups, the *Monachidae* and the *Phocinae*, are distinguished from otariids by the facts that they do not have ear flaps, they swim with their feet, they have a big blubber layer and relatively large ear bones. They are known from the upper Oligocene. A fossil, found in South Carolina is, in fact, the oldest fossil associated with pinnipeds to date. Apart from this discovery, the other fossils are more recent, dating from middle Miocene.

Acrophoca longirostris, primitive seal. Smithsonian Institution.

Allodesmus.

Allodesmus kernensis.

MOUNTING MARINE MAMMALS SKELETONS

Mounting marine mammals skeletons does not require techniques very different from those used in mounting other vertebrate skeletons, with the exception, maybe, of the very large cetaceans. And, even in that case, it is not much more complicated than mounting a big dinosaur.

The first thing to do is to flense the carcass. In the case of small cetaceans and most pinnipeds, it is not a major problem. The meat can be boiled away in a cauldron or with water and steam under pressure, or one can let mother nature, meaning the fly larvae, do the work. The last method takes more time, an isolated place so the putrefaction odors won't bother the neighbors, and finally a strong heart, to verify the state of decomposition and take away the water and juices that accumulate in the container in which the carcass has to be placed so scavengers won't drag it away. If this is not done, the maggots and other insects will drown and part of the bones won't be cleansed. It is messy but by far the less expensive method.

Maggots at work on a seal skeleton.

3 weeks later.

Once the bones are free from the meat, tendons, and ligaments, the remaining grease has to be removed. If it is allowed to remain, it will drip out from them and take on a rancid odor. The dust will stick on skeleton and various molds can develop on the greasy parts of the bones. It is, in fact, one of the major problems one faces when mounting a big cetacean. You can see that this has not always been done properly in museums where a big whale's skeleton are exhibited.

It can be solved in different ways. The bones can be placed in large containers full of chemical degreasers, like acetone or several others, more potent but more carcinogenic. Huge quantities are needed, which makes the work doubly expensive because one has to pay for the usually very costly chemicals and pay again for specialized enterprises to dispose

Blue whale skeleton. The grease gives a yellowish tint to the bones and it will melt to the lower parts. *New Bedford Whaling Museum.*

of the by-products, which usually are hazardous to the health. Another method consists of putting the bones in big containers filled with water solutions of special enzymes (lipases) which are supposed to act on fat only and not on the other organic materials that give the bones their strength. These products are expensive and require some work while they act (agitation of the water, maintaining a constant temperature). In one final, but slow method, one can just put plain water in the containers with the bones, and maintain its temperature in the vicinity of body temperature. It is called maceration. Bacteria will develop in the water, and act on what flesh may remain, and the denser water will penetrate the bones and push out the grease. The nearer of the body temperature the water is, the faster the work will be done because fat will be more fluid. For a big fatty skeleton it can take several months. The great advantage is that no chemical products (costly and risky) are used. The disadvantages are the time and the smell. Those can be controlled to a degree by changing the water often enough.

I do not recommend this method, but in countries with temperate climate and cold winters the carcass can be buried in manure or the ground. If the animal is small enough you will have to hire some archeologists to exhume the bones if you want to be sure that you did not lose any small bone and they will be colored almost indelibly by components in the manure or ground. If the animal is big enough the risk that bones will be broken during burial and exhumation are great, and it is very difficult not to lose anything small. I have once, full of enthusiasm (it was my first big one), tried to exhume a Fin whale buried 13 years before on a beach. All the blubber was still there and most of the vertebrae processes had been broken. Lots of work for finally being able to mount only the head. It must be added that the soil minerals will stain the bones in a way almost impossible to clean.

Once the skeleton degreased, it can be bleached. Hydrogen peroxide is very good for that. It is more economical to buy it at very high concentrations and dilute it later to 30%. Be careful in using it. It is a powerful oxidizing agent and it can cause severe burns. It is possible, too, if you are not in a highly polluted environment (full of industrial dust) and not in a hurry, just let the sun bleach your bones for one or two weeks. To put bones of large cetaceans in the hot sun also helps with the degreasing. Heat melts away the fat from the bones and it will drip out of them. Never use bleach; it has a very bad effect on bones leaving them chalky and flaking.

The next step will be the classification of the bones according to their anatomical place. In most of the cases it is easy if the bones from the front limbs have been separated from those of hind limbs (in pinnipeds at least). In certain cases, it is difficult without experience (or a very good book) to distinguish the phalanges of the hands from those of the feet.

Bones from a Minke bleaching under the sun.

When bones articulate firmly, one with the other, the shape of the joint of one determines the shape of the joint of the other, and some attention and a certain sense of mechanics will allow an easy assembly. It becomes a little more complicated when there is no contact between adjacent bones, like in the hands of cetaceans for example. In this case the bones of each structure must be separated before preparation. They can be drilled and fixed on some kind of a board, so they will stay in place through the cleaning process. You can drill them and put a number on each, or make an X-ray picture of the structure before cleaning. In some cases (small cetaceans) one can try to maintain the ligaments intact between the bones. Then, when the prepared structure dries, the ligaments act as glue and if the shrinking is not too important the mount is quite faithful to the original.

Sorting the ribs is not complicated because their shape and their length will allow them to be separated into left and right groups easily enough.

Partially classified seal skeleton.

X ray picture of a Beluga limb.

Classification of a dolphin bones.

Vertebrae are easily sorted too. Mammal's cervical vertebrae (except the last one) display a hole on each side of the neural canal for the vertebral artery to pass through. The last one is almost like the first thoracic but lacks the articular facets for the ribs.

Cetaceans cervicals are either welded as in most Odontocetes and Right whales, or very thin when compared to the others. Tranverse processes are vertical.

Thoracic vertebrae are easily identified by the articular facets they possess to articulate the ribs which are attached to them. Almost all the ribs are attached at 2 points, except in the cetaceans (only the first 5 in Odontocetes and the first 2 in Mysticetes).

Grey seal, 6th cervical.

Cervical vertebrae Mysticete (A); Odontocete (B).

Thoracic vertebrae: pinniped (A); cetacean (B).

Lumbar vertebrae do not have holes on the side nor articular facets for ribs. There are a maximum of 6 in pinnipeds but a lot more in cetaceans (11-24) There are no articular facets for ribs, but processes that limit movements (pre- and post-zygapohyses)

Sacral vertebrae are welded in one bone, the sacrum. They do not exist in cetaceans in which there remains only a vestigial hind limb, but they are found in pinnipeds.

Cetaceans caudal vertebrae are recognized by the pair of articular facets where the Y bones, the chevrons, articulate.

Lumbar vertebrae pinniped (A); cetacean (B).

There are 2 pairs of those facets 2 anterior (cranials) and 2 posterior (caudals) on the ventral side of each vertebra. The chevrons are attached to the caudal facets. The last caudal vertebrae do not support chevrons.

Seal sacrum.

Caudal and lumbar vertebrae, ventral view, comparison.

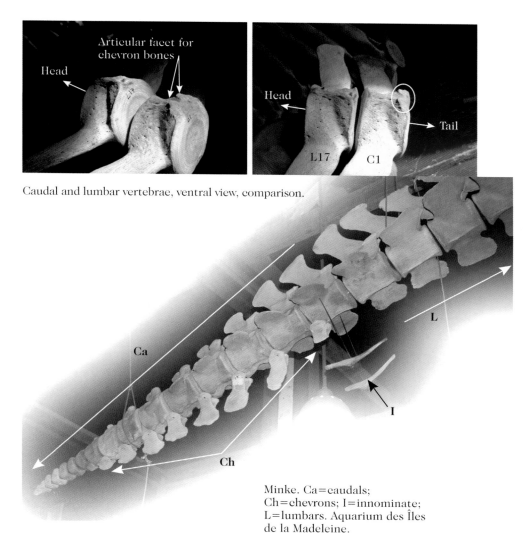

Minke. Ca=caudals; Ch=chevrons; I=innominate; L=lumbars. Aquarium des Îles de la Madeleine.

277

Cetacean ligamentous mounting.

Pinniped front limb.

Cetacean non-ligamentous mounting.

Pinniped hind limb.

The cetaceans have only one pair of limbs, pinnipeds two.

A limb is made of several parts. The girdle which attaches it to the trunk, the arm or the thigh, the forearm or the leg, and the hand or the foot.

One has to be very careful when collecting the bones of a specimen, so not to lose any bones. Some can be very tiny, like phalanges, others can be small and not connected to the rest of the skeleton, like the innominates in cetaceans which are the remnants of the pelvic girdle of their 4-limbed ancestors. They are not difficult to find once you know where to look for them. They are on each side of the genital slit, rather superficially placed, almost right under the blubber.

Innominate, Fin whale, female. Ste Flavie, 2001.

Innominate, Blue whale female. 7 Îles 2004.

Innominate, Blue whale male. Anticosti, 1987.

Innominate Minke male. Îles de la Madeleine, 2003.

Mounting

You can use different methods but all must respect the anatomical position of the bones in the living animal. The entire skeleton can be mounted and suspended or placed on any kind of stand, or only half of it can be put on a board. This last method implies that you will have to sacrifice half of the skeleton or at least the vertebrae, because the spinous process is too thin to be cut in half. Only in pinnipeds is it thick enough so two half skeletons can be mounted with one specimen.

This method allow the skeleton to be placed on a wall, it takes less space , and if one has to carry the specimen for expositions or lectures, it is more easily packed, and less fragile. I have a seal and a porpoise skeletons that have traveled thousands of km with me for expositions or conferences without sustaining any major damage.

Complete Skeleton.

1/2 Skeleton.

ERRORS MADE IN MOUNTING CETACEANS SKELETONS

When you have a skeleton to mount you need to look to what has been made before by specialists either by studying books or by going to museums. Most of the time the method is good but in some cases, as with the rorqual skeletons, it could lead you astray. All the rorqual skeletons I have seen in several museums (and big, illustrious ones!) have been badly mounted, and this since the beginning, I think. The errors are repeated because the new specimens are mounted as were the ones made in the 19th century or even before.

CAUSES OF THE ERRORS

The first cetaceans skeletons to be mounted were those of the Right whales, since only those were hunted. In late 19th century two very good cetologists, Gervais and Van Beneden, published a very beautiful book, *L'ostéographie des cétacés vivants et fossiles*. As long as one looks at the individual bones represented it is perfect, almost better than pictures. It is not as good when complete skeletons are represented: the ribs are drawn vertically, as hanging from the vertebrae, the lower jaws are flat and very near the rear part of the skull, and hyoïd bones and innominates are rarely in their anatomical position.

Minke skeleton, Osteographie Gervais and VanBenede.

This book, considering the quality of its drawings, is a reference, and it is quite sure that any technician to whom is given the heavy task (pun not intended) of mounting a big whale skeleton will use it as a guide, then will go to different museums, taking pictures. (It is rarely the same person who collects the bones and mounts the skeleton.)

I did exactly this when I had to mount my first Fin whale for a thematic pavilion for the 1534-1984 festivities in Québec City. I went to museums, I looked at the Gervais van Beneden book and so I repeated the same errors! The lower jaws were wrong, and so were the mandibles and the hyoïd bones.

Fin whale, H2O pavilion, Québec 1984.

Time went by and I started to collect cetaceans skeletons and did a lot of necropsies on several species of rorquals; some details became apparent.

Contrary to Right whale jaws, it is impossible to put the mandibles of a rorqual in a horizontal position when the mouth is closed, if we want its mouth to close!

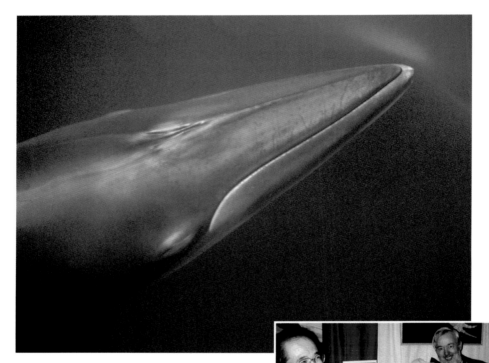

Minke alive, the mandible seems straight.
U.Tscherter (ORES)

"Pierre's Position." From left to right Dr. Tadasu Yamada, Dr. Joy Reidenberg, Dr. James G. Mead.

In fact, the mandibles of the rorquals are curved, and if they are placed flat, the curve goes outside. The lip should be quite longer, so that it is to be able to close the gap between upper and lower jaw. When one looks at a living rorqual mouth closed, it is evident that the mandible is parallel to the maxillary. The only way it can be so is to rotate the mandible so the curve is vertical. Lambertsen has given a good description of the rotation of the mandible when the mouth opens or closes. During the 14th Biennial on Marine Mammal Biology (2001) I shared my point of view with Drs. J.G. Mead (Smithsonian),

Correct position of the mandibles.

J.Reidenberg (Mount Sinai), and Tadasu Yamada (Tokyo), and they all agreed (after manipulating for a long time a Minke fetus skull I had with me) that my hypothesis on the way mandibles should be placed in mounted, closed mouth rorqual skeletons was correct.

In most rorquals mounts, the mandibles are way too much toward the rear of the skull. In the living animal they are separated from the skull by a fibrous cushion (see the chapter on nutrition). This should be taken into account when mounting the skull.

Only Right whales (and maybe the Grey whales) have a synovial joint and the mandible with the curve in the horizontal plane. But they have huge lips to help hold their immense baleen plates (2-4,5m).

Right whale. Galerie d'Anatomie Comparée, Paris.

As far as ribs are concerned, if they are placed as in Gervais's book, the flat part is oriented towards the head, and the thin one on the side, but, in reality, the flat part is always oriented toward the exterior of the body.

The hyoïd apparatus, situated at the base of the tongue, is placed in many ways, but I have seen it only once almost correctly mounted.

It is attached all the way to the rear of the skull and in the living animal very near to the skull, almost under the cervical vertebrae. The position of the apparatus very near the skull may be a reason why the baleen whale cannot swallow big prey: it cannot move down and the space between it and the skull is way too small for big prey to pass through.

283

The ribs of a Minke. It is evident that the flat side is toward the exterior.

Minke, stylohyal in situ.

Minke, hyoïd apparatus correctly placed, Iles de la Madeleine 2004.

The innominates, those remnants of the pelvic girdle of whale's terrestrial ancestors, are almost always placed too near the vertebrae. Those bones are just under the blubber, on each side of the genital slit. To place them correctly, one has to imagine the shape of the body near the genital slit, just a little forward of the first caudal, and place them almost as far from the vertebrae as the skin of the belly would be.

I do not know any museums, except those named below, in which there is a correctly mounted rorqual skeleton. I have visited lots of museums, I have seen pictures of more, and all the skeletons present the defect mentioned, in part or in totality (and in some there are even more errors!).

I can understand a curator who tells me that he has no money to change the mount. What I do not understand is why there is not a notice explaining what is wrong and the reasons why it remains so. A museum is a place where the main preoccupation should be a pedagogical one. To make an error is possible, but not wanting to correct it denotes a total lack of the sense of responsibility that an institution of high quality should possess when passing on information. It only succeeds in perpetuating errors, if a novice, as I once was, is inspired by the work exhibited. It is interesting to mention, though, that people visiting the new Natural Sciences Museum in Toulouse, France, when it opens its doors, will see a correctly mounted Fin whale skeleton. And in Quebec province, good Minke whale skeletons can be seen in Aquarium des Îles, Iles de la Madeleine, the CIMM Museum at Tadoussac and in Le Musée du Squelette on Île-Verte.

Some examples of badly mounted skeletons The name of the museum is not given because we want only to show what tradition can do. Arrows point the classical errors.

Techniques

Mounting skeletons can be done with glue, metallic wires turned spring-like at each end, and with screws and bolts for the bigger specimens.

Vertebrae will be placed on a rod strong enough for the size and weight of the specimen. A way of preventing the vertebrae from rotating around the rod has to be found. A square tubing can be used, vertebrae being screwed to the tubing, or two superposed rods can be used. In fact, one has to adjust to the various specimens and situations. The visibility of the frame should not be a big worry, as long as the metal used is not too wide or thick (steel is better than aluminum, because its strength and elasticity allow smaller gage to be used). When someone is looking at a whale skeleton he forgets, very soon, the frame and concentrates on the bones.

If the skeleton has to be suspended, hanging devices will have to be installed.

Some steps in mounting a skeleton.

287

Beluga, Île Verte.

Tursiops, Grey seal, White sided dolphin, Musée du Squelette, Île verte.

A footprint on the sea... 80 tons of whale have just passed!

Fact sheets

THE CETACEA

THE MYSTICETES

The Cetacea

The order Cetacea is divided into 2 sub-orders:

A. Mysticetes, divided into 4 families, 5 genera, and 10 species;

B. Odontocetes, divided into 6 families, 35 genera, and 70 species.

    The facts sheets contained within this book do not cover all the known species of the world. Only the most familiar whales and those that frequently strand on the shores of Eastern North America are included.
    The main objective of this book is not to provide exhaustive fact sheets, but to give as much information as possible on the biology and ecology of cetaceans.
    Many excellent works that deal almost exclusively with cataloging cetaceans of the world are available; some are given in the reference section of this book.

The Mysticetes

Etymology: Greek: *mystakos*: mustache; Latin: *cetus*: whale

Mysticetes (baleen or whalebone whales) have two blowholes.

Teeth are absent in adults, but vestigial in fetuses. These whales are characterized by specialized structures of dermal origin called BALEEN used to strain plankton from the water. There are two methods of sieving food, based on the size of the baleen, the head and the buccal cavity. There are consequently two groups of Mysticetes.

A. Skimmers: With their mouth open, they swim close to the surface in the clouds of animalcules on which they feed, continuously filtering the seawater. The baleen plates are long (from 2 to 4.5 m; 6.6 to 14.8 ft) and narrow. The head is very large, up to one-third the total length of the body; the rostrum is considerably narrow and arches upward. Throat and ventral grooves are absent.
B. Swallowers or gulpers: They feed by gulping single mouthfuls of water containing food (small crustaceans or small fish). Once the mouth is closed, the water is forced out through the baleen by the contraction of throat muscles and the tongue. Prey items are trapped on the inner surfaces of the whalebone inside the mouth. The head is smaller, the baleen is shorter (maximum of 1 m; 3.3 ft), and the rostrum is wide and flat.

These whales have many folds or grooves on the throat and belly that allow the buccal floor to balloon out as water enters their mouth. Depending on its size, a Blue whale can take in from 20,000 to 35,000 liters of krill-filled water in a single mouthful.

Mysticetes are grouped into four families:

- *Balaenidae*, from the Latin balaena: *whale*
- *Balaenopteridae*, from the Latin *balaena*: whale and Greek *pteron*: wing or fin.
- *Eschrichtiidae*, named for Eschricht, a nineteenth-century Danish naturalist.
- *Neobalaenidae* (one species only) *CAPEREA MARGINATA* (the Pygmy Right whale)

THE BOWHEAD WHALE OR GREENLAND RIGHT WHALE

Balaena mysticetus

ORDER Cetacea
SUBORDER Mysticeti
FAMILY Balaenidae (Latin: whale)
SPECIES *Eubalaena mysticetus*
Greek *eu*: good and *mystakos*: mustache
Latin *cetus*: whale

VERNACULAR NAMES Greenland Right whale, Polar whale. These excessively fat animals were indeed the "right" whales to hunt because they floated when killed. (Fr. Baleine franche du Groenland.)

DESCRIPTION

Length: On average, 15 m (49 ft). Maximum: 20 m (66 ft)
Weight: 60 to 100 tons
Color: Black, blackish-blue, sometimes very dark brown. A light greyish band is occasionally seen on the caudal peduncle (or tail stock). It bears a characteristic white patch on its chin. The lack of facial callosities distinguishes it from the Black Right whale (*Balaena glacialis*).
Dorsal fin: Absent
Caudal fin: Up to 8.5 m (~28 ft) in width
Head: At least one-third of the total body length, with a strongly arched rostrum and very high lower lips.
Baleen: 230 to 360 plates on each side (330 on average) length: Up to 4.5 m (14.8 ft) width: A maximum of 36 cm (14 in.) Color: Dark grey
Ventral grooves: Absent
Skeleton: 53 to 55 vertebrae, the 7 cervical vertebrae are generally fused together (at least in the adult). The hand has 5 fingers.

A Right whale skeleton as seen by the artist Caroline Monette.

NATURAL HISTORY

Breathing sequence: 4 to 6 breaths every 40 seconds or so

Dives: 20 minutes with maximums of 40 minutes. It usually performs shallow dives and shows it flukes as it dives.

Blow: A very distinct V-shaped spout reaching 7 m (23 ft) high. There is a wide separation between the two blowhole openings.

Diet: Adapted to feed on microplankton. The whalebone apparatus formed by the very fine fringes of the baleen plates allows it to eat to krill and copepods such as *Calanus plumchrus* and *Calanus cristatus*. Its blubber layer can reach a thickness of 70 cm (2.3 ft).

Vocalizations: 160 to 500 hertz

Reproduction: Gestation lasts from 12 to 14 months (research estimates vary). Calving occurs mostly between April and May. Nursing lasts approximately one year, by which time the calf has then doubled its length. At birth, it measures between 3.5 and 4.5 m (11.5 and 14.8 ft) long.

Longevity: Probably more than 200 years. In fact, in May 1993, in the carcass of a very large bowhead (killed by Inuits), two stone harpoon points were found, one of which was last known to be in use in the early 1880s and not later than 1900. This longevity has been confirmed by the method described on page 181.

Population: The worldwide population is estimated at 7,500.

Distribution: There are 4 main populations – one in the western part of the Arctic, one in the Sea of Okhotsk, another around Baffin Bay and in Hudson Bay, and finally, one from Greenland to the Barents Sea, which is virtually extinct.

Behavior: Slow swimmers (5 km/h, 2 to 3 kn., or 3.1 mi./h) that can travel in groups of 6 to 60 individuals. It is a migratory species.

THE BLACK RIGHT WHALE

Eubalaena glacialis

ORDER	Cetacea
SUBORDER	Mysticeti
FAMILY	Balaenidae (Latin: whale)
SPECIES	*Balaena glacialis* Latin: *balaena*: whale and *glacialis*: ice
VERNACULAR NAMES	Northern Right whale, Biscay Right whale, (these excessively fat animals were indeed the "right" whales to hunt because they floated once they were killed). (Fr. Baleine franche noire.)

DESCRIPTION
Length: On average, 15 m (49 ft). Maximum: 18 m (59 ft)
Weight: 30 to 100 tons
Color: Usually black with occasional brown patches.

There are white patches whose numbers vary, on the ventral side. The head has outgrowths or callosities (keratinized skin patches) that are infested by barnacles, whale-lice, and other organisms. The callosities are found on the top of the head ("bonnet"), on top of the eyes ("eyebrows"), in blowholes, and under the chin. Calling them callosities is misleading—it suggests that they are outgrowths that form over worn-out skin. But this is not the case since fetuses already have them.

Dorsal fin: Absent
Caudal fin: 7.5 m (~25 ft). Deeply notched. The posterior rim is concave.
Head: Huge, more than one-fourth the total body length. The rostrum is strongly arched.
Baleen: Not as long as the baleen plates of the bowhead, but measures close to 2 m (6.6 ft). There are from 205 to 270 plates on each side. They are light grey and seem yellowish when the whale feeds at the surface of the water.
Ventral grooves: Absent
Skeleton: Skull with a very narrow and arched rostrum. There are 14 to 15 pairs of ribs, 56 or 57 vertebrae; the 7 neck bones are usually fused in the adult. There are 5 fingers in the hand.

NATURAL HISTORY

Breathing sequence: 12 to 15 breaths every minute or so, between dives
Dives: Normally 6 to 8 minutes. Longest recorded dive is 60 minutes. Shows its tail as it dives.
Blow: The very distinctive V-shaped blow is up to 5 m high (16 ft). The two blowhole openings are widely separated.
Diet: It feeds on copepods (*Calanus sp.*) and euphausids larvae (krill) by slowly swimming through clouds of animalcules, at the surface or at greater depths under water.
Vocalizations: Moans, groans, and various pulsed sounds ranging from 30 to 2,200 hertz.
Reproduction: Gestation lasts one year. Nursing lasts for about 10 months, but the calf stays with its mother a little while longer. Breeding and calving take place from December to March in the Northern Hemisphere. Sexual maturity is reached when the animal measures about 15 m (49 ft),

probably at the age of ten. Copulation is done while both partners are on their side, following a rather lengthy courtship period. There is no violence among males (at least in the Southern Hemisphere). Females give birth every three years. Newborns measure 5 to 6 m (16 to 20 ft) in length. Right whale's testicles, having a combined weight of 1,000 kg (2,205 lbs.) during the rutting seasons and are the largest of any of the bigger cetaceans.

Longevity: Its life span is similar to that of the Bowhead whale, probably more than 200 years. A sexually mature and distinctively marked female was photographed in 1935 with a calf, so she was at least eight to ten years of age at the time. She has been sighted several times since, most recently in 1994. The method described on page 181 has confirmed that Black Right whales can live more than 200 years.

Population: 870 to 1,700 in the Northern Hemisphere, 8,000 in the Southern Hemisphere. It is a species whose long-term survival is highly uncertain in the Northern Hemisphere.

Distribution: Northern Hemisphere: Along the east and west coasts of North America, from Labrador to Florida and Alaska to southern California. Along the east coast of Asia. Southern Hemisphere: On the east and west coasts of South America and Africa, and from southern Australia to New Zealand.

Behavior: Slow but very acrobatic swimmers. Breaching, pec and tail slappings seem to be done for no apparent reason. They live in small groups.

Skeleton of a Right whale. Compare the size of the person at the left with the size of the baleen.

Musée du Jardin des plantes, à Paris

Williamson

The distinctive bonnet, a protrusion on the head in front of the blowhole, is used to identify individuals.

New England Aquarium

Demetri Marguérat

A northern Right whale spy hopping.

A sounding northern Right whale showing its tail.

THE BLUE WHALE

Balaenoptera musculus

ORDER	Cetacea
SUBORDER	Mysticeti
FAMILY	Balaenopteridae
SPECIES	*Balaenoptera musculus* (two subspecies, *Balaenoptera musculus intermedia* and *Balaenoptera musculus brevicauda* in the Southern Hemisphere)
VERNACULAR NAMES	Blue rorqual (Old Norse: rør grooves; hval-whale) and Sulphur-bottom for the diatoms, *Cocconeis ceticola,* or yellowish algae that fix themselves on the skin of its belly. (Fr.: Baleine bleue or Rorqual bleu.)

DESCRIPTION

Length: The largest "scientifically"-recorded specimen was a 33.58-m (110.17-ft) female (i.e., measured from the tip of its chin to the notch of its tail), found in southern Georgia between 1904 and 1920. Here, in the St. Lawrence River, the longest specimen that I have measured was a 22.4-m (73.5-ft) female (Anse-aux-Fraises, île d'Anticosti, 1985). An adult male that beached on the shores of Jupiter-la-mer, île d'Anticosti, and "Pita," a male of at least 13 years of age, both measured between 21 and 22 m (69 and 72 ft). "Pita," which had been photographed on numerous occasions by Richard Sears, was found dead at Chicotte-la-mer, île d'Anticosti, in 1992.

Weight: The record weight was established in southern Georgia by a female measuring 27.6 m (90.6 ft) that was killed on March 27, 1947. She weighed an estimated 190 tons (Tomiline, 1957). Another female, killed in southern Georgia in 1931, measured 29.5 m (96.8 ft) and is said to have weighed 174 tons (Laurie, 1933). The average weight of a Blue whale ranges from 75 to 130 tons.

Color: Greyish-blue, slate-grey with lighter patches. Each whale has its own distinctive pattern, used in photo-identification. The underside of the pectorals is whitish.

Dorsal fin: Small, about 30 cm (12 in.), located at one-fourth the body length from the notch of the tail. Falciform or triangular in shape.

Caudal fin: 6.5 to 7.5 m (21.3 to 25 ft). Posterior rim rectilinear. Deeply notched.

Head: U-shaped when seen from above. Its splashguard or blowhole crest is particularly visible when it breathes at the surface.

Baleen: Bluish-black, 270 to 395 plates on each side of the mouth measuring no more than 1 m (3.3 ft), often wider than long.

Ventral grooves: 55 to 88 grooves, the longest starting from the chin and ending at the navel. They allow the floor of the mouth to balloon out, increasing its volume sixfold; the mouth can consequently take in 25,000 to 35,000 L of water.

Skeleton: Skull with curved maxillaries (forming a U-shaped rostrum, seen from above). There are usually 63 to 64 unfused vertebrae; some may fuse in older individuals ("Pita's" second and third cervical vertebrae were fused together). There are only 4 fingers in the hand.

NATURAL HISTORY

Breathing sequence: 8 to 15 breaths (at 15 to 20 second intervals) between dives when the animal is active

Dives: Normally 10 to 30 minutes. Will occasionally show its tail as it dives.

Blow: 6 to 12 m (19 to 39 ft)

Diet: This whale feeds almost exclusively on krill (*Thysanoessa sp.* and *Meganyctiphanes norvegica*) by engulfing huge amounts of krill-laden waters. It either feeds near the surface or deeper under water. It also feeds on the copepod Temora longicornis. At its summer feeding grounds, it consumes about four tons of food daily (a 120-day period).

Vocalizations: Moans, groans, and various pulsed sound. Ultrasounds, emitted by clicks, could represent a type of biosonar in Mysticetes (Beamish and Mitchell, 1971).

Reproduction: Gestation lasts one year. Breeding and calving take place toward the end of fall and winter in the Northern Hemisphere. Sexual maturity is reached when the animal measures about 20 m (66 ft), probably at the age of ten. A female usually gives birth every three years, with normally one calf. A calf measures 5 to 7 m (16 to 23 ft; 2.5 tons) at birth and 16 m (52 ft) when weaned seven months later. It grows at rate of 3.7 kg/h (8.2 lbs./h) during its nursing period!

Longevity: More than 100 years

Population: About 9,000 worldwide. Its long-term survival is extremely problematic. Numbers prior to whaling were around 200,000.

Distribution: In all the world's cold waters

Behavior: Solitary or in small groups (probably because of high food concentration). It normally swims at speeds of 6 to 8 km/h (3.7 to 5 mi./h), but can exhibit extraordinary bursts of speed when disturbed (40 km/h, 25 mi./h). It is a migratory species, moving from its colder feeding grounds to the equatorial temperate waters during winter.

Color pattern. The dorsal fin always seems smaller.

Only a full scale model allows us to realize the gigantic size of a Blue whale. This one can be seen in American Museum of Natural History, New York.

ORES - Ursula Tscherter

The blow of a Blue whale is always very high.

The caudal fin is not always seen out of the water as the animal sounds. The tail flukes are much more massive than that of the Humpback whale.

The muscle contraction during the opening of the blowholes forms an important mass.

The scars on the back of this Blue whale permit scientists to identify it.

Pierre-Michel Fontaine

BREATHING SEQUENCE OF THE BLUE WHALE

THE FIN WHALE

Balaenoptera physalus

ORDER	Cetacea
SUBORDER	Mysticeti
FAMILY	Balaenopteridae
SPECIES	*Balaenoptera physalus* (Greek *physalus*: bellows)
VERNACULAR NAMES	Finback whale, Finner, Razor Back, and Flathead. (Fr. Rorqual commun.)

DESCRIPTION
Length: In the Northern Hemisphere, males are 22 m (72 ft) and females, 24 m (79 ft). In the Southern Hemisphere, males are 25 m (82 ft) and females, 27 m (89 ft)

Weight:	C. Lockyeyr (1976) has established a relationship between body mass and length from data collected during commercial whaling. According to the curve obtained, a 25 m (82 ft) Fin whale would weigh 85 tons whereas one which measures 20 m (66 ft) would weigh 50 tons. For whales up to 20 m (66 ft) in length, this curve is practically identical to that of Blue whales.
Color:	The back is dark brown or dark grey; the belly and the underside of the pectorals are white. The lower right jaw is white. Light brown or grey stripes called chevrons originate behind the blowhole and run aft to form a broad V along the back of the animal.
Dorsal fin:	Various falciform or triangular shapes. It is relatively large, about 60 cm (2 ft) and is located over the posterior two-thirds of the body.
Caudal fin:	5 to 6 m (16 to 20 ft) wide. Posterior rim is lightly concave. Deeply notched.
Head:	V-shaped when seen from above.
Baleen:	Greyish-brown, 260 to 480 plates on each side of the mouth. On the right side, 20 to 30% of the baleen are white or light yellow.
Ventral grooves:	56 to 100, the longest starting from the chin and ending at the navel. It allows the floor of the mouth to balloon out, considerably increasing its volume. The mouth can then take in close to 20,000 L of water.
Skeleton:	Skull with pointed maxillaries (forming a V-shaped rostrum, seen from above). There are usually 60 or 65 unfused vertebrae; some may fuse in older individuals. One very huge female that washed up on the shores of Lac Saint-Pierre (near Trois-Rivières, Québec) had many of its lumbar vertebrae fused together. The hand has four fingers (the middle finger is missing, as is the case with all rorquals).

NATURAL HISTORY

Breathing sequence:	6 or 7 breaths between dives (at 15 to 20 second intervals) when the animal is active
Dives:	Normally 10 to 20 minutes at depths of up to 250 m (820 ft). It does not show the tail as it dives.
Blow:	4 to 6 m (13 to 20 ft)
Diet:	This whale feeds almost exclusively on krill (*Thysanoessa sp.* and *Meganyctiphanes norvegica*).

But it is also an opportunistic feeder, eating capelin, sandlance, young herring, and squid, according to the time of year and its feeding grounds. At its summer feeding grounds, it consumes about three tons of food daily (a 120-day period).

Vocalizations: Moans, groans, and various pulsed sound ranging from 20 to 100 hertz

Reproduction: Gestation lasts one year. In the Northern Hemisphere, breeding and calving usually take place between December and January. Sexual maturity is attained at about 18 m (59 ft) for females, and 17 m (56 ft) for males, that is, at the age of ten (values established at the start of intensive whaling). With the subsequent rarifaction in the Fin whale stock, the age associated with sexual maturity is now six. Usually one calf every three years. Females give birth every three years, normally to one calf. Newborns measure 6 m (20 ft; 1.9 tons) and 12 m (39 ft) when weaned six or seven months later. Fifty years of whaling also took its toll on the breeding cycle. According to Laws (1961), the proportion of nonlactating gravid females reached 50 to 55% which represents half the normal intervals between gestation periods.

Longevity: More than 100 years. According to data collected by the Japanese and cited by Mizroch, one year of age in the Fin whale is equivalent to two layers of wax (one light and one dark) in the EARPLUG. Such layers are collectively called growth layer groups.
The oldest specimen, captured near Antarctica, was 111 years old.

Population: About 123,000 worldwide. Its long-term survival is highly uncertain. Numbers prior to whaling were around 300,000 to 650,000.

Distribution: In all the world's cold waters

Behavior: Solitary or in small groups (probably because of high food concentration rather than a particular social structure). It normally swims at speeds of 6 to 8 km/h (3.7 to 5 mi./h), but can exhibit extraordinary bursts of speed when disturbed (40 km/h, 25 mi./h). It is a migratory species, moving from its colder feeding grounds to the equatorial temperate waters during winter).

"Boomerang".

"Capitaine Crochet," August 26, 1996.

"Perroquet"

Diatoms on the sides of Fin whales.

"Bossu".

A lazy Fin whale.

Fin whales near île Verte.

"Razor Back"

The scar on this Fin whale's dorsal fin will allow us to identify it.

311

BREATHING SEQUENCE OF THE FIN WHALE

On very rare occasions, the Fin whale will actually show its tail as it dives. This rare happening was seen off Gaspé (Québec) in August 1997. Notice that a fluke is missing, probably accounting for this unusual spectacle.

THE MINKE WHALE

Balaenoptera acutorostrata

ORDER	Cetacea
SUBORDER	Mysticeti
FAMILY	Balaenopteridae
SPECIES	*Balaenoptera acutorostrata* (Latin *acutus*: sharp or point; *rostrum*: snout)
VERNACULAR NAMES	Lesser rorqual, Meineke's whale, Piked whale, Sharp-nosed whale, and Bay-whales. Two stories about a German whaler named Meineke give reasonable accounts for the origin of the name. First, Meineke may have mistaken this relatively small whale for a Blue whale. According to the other version, Meineke was in the habit of overestimating the size of the whales killed by his boat. In mockery, the other whalers called this whale Meineke's whale. (Fr. Petit rorqual, Gibard.)

DESCRIPTION

Length:	Rarely more than 10.5 m (34 ft). On average, 8 to 9 m (26 to 29 ft). Maximum 12 m (39 ft)
Weight:	6 to 8 tons
Color:	The back is dark grey, black or brown. In the Northern Hemisphere, it has a white band on its pectoral fins, distinguishing it from all the other rorquals. This marking is not constant in the Minke whales of the Southern Hemisphere.
Dorsal fin:	Located at the posterior two-thirds of the body. It is relatively large and strongly curved or hook-shaped. It appears at the same time as the spout does.
Head:	Small and pointed rostrum
Baleen:	Yellowish, packed and finely frayed (297 plates on the right side of the mouth of a stranded Minke whale near Ragueneau, Québec, June 1995). On the right side, 20 to 30% of the baleen are white or light yellow.
Ventral grooves:	60 to 70 grooves on the throat and chest but do not reach the belly.

NATURAL HISTORY

Breathing sequence:	5 to 8 breaths at 30-second intervals.
Dives:	It hunches up its back on its last breath prior to diving, does not show its tail.
Blow:	Up to 2 m (~7 ft).
Diet:	Krill, young herring, sandlance, and capelin. In the stomach of a beached Minke (near the mouth of Bersimis River, Québec. September 1992) fragments of sandlance (identified by their otoliths) were found.
Vocalizations:	Groans, moans, and usually low-frequency metallic sound, but also may reach 100 Hz to 12 kHz.
Reproduction:	Breeding occurs from December to March in the Northern Atlantic. Gestation lasts 10 to 11 months. Calving usually takes place from November to March. Newborns measure 2.4 to 2.7 m (7.9 to 8.9 ft); 350 kg (772 lbs.). Calves are weaned six months later. Sexual maturity, in females, is attained at about 7.3 m (24 ft), at the age of 7.1; in males, sexual maturity is attained at 6.7 to 7 m (22 to 23 ft) at the age of 6 (estimated from growth layers on the tympanic bulla).

Longevity: Close to 50 years
Population: About 900,000 worldwide
Distribution: Worldwide. Frequent in coastal areas and estuaries
Behavior: Solitary or in small groups of two or three individuals. May form pods. Speeds: 24 to 30 km/h (15 to 19 mi./h). Fast swimmers and very curious, they often approach ships. I once saw them ride the waves created by a moving boat as I was coming back from a whale-watching tour at Baie-Sainte-Catherine, Québec. They are unfortunately attracted to fishing nets. The Minke whale performs a series of breaches. When it feeds, it can heave half its body out of the water with its throat distended by the water just taken in. One of its favorite feeding techniques consists in rushing in a school of fish from below. It sometimes turns on its sides to swallow its prey.

A stranded Minke whale on the shores of the Gaspé peninsula, July 1997.

The fingers point to the opening of auditory canal (on the left), and the eye (on the right). Petits-Escoumins, October 19, 1997.

ORES - Ursula Tscherter   Breaching of Minke whales.   *ORES - Ursula Tscherter*

No, it is not an Orca's dorsal fin, but half of the fluke of a Minke swimming on its side while feeding.

The white band on the pectoral fin is clearly visible on this picture.

A Minke whale feeding.

Even in murky waters, it is possible to see the white band on the pectoral fin.

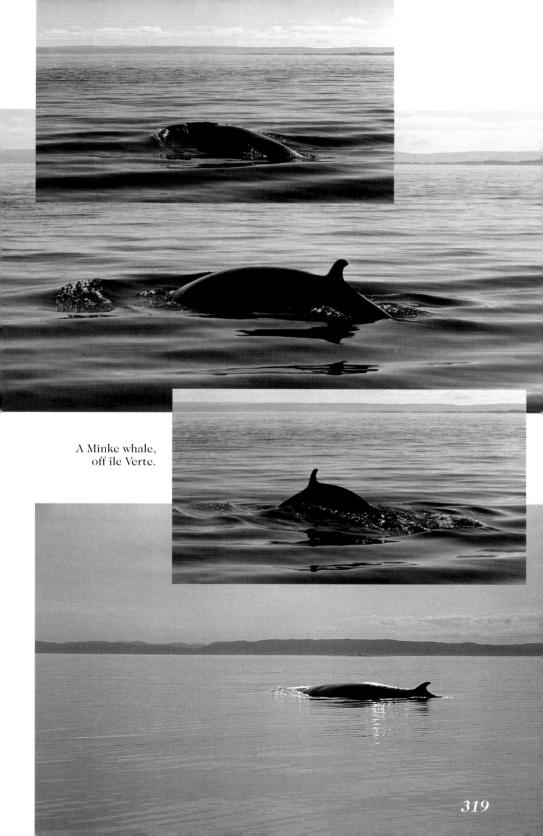

A Minke whale,
off île Verte.

BREATHING SEQUENCE OF THE MINKE WHALE

THE HUMPBACK WHALE

Megaptera novaeangliae

ORDER	Cetacea
SUBORDER	Mysticeti
FAMILY	Balaenopteridae
SPECIES	*Megaptera novaeangliae* (Latin: *mega*: big; *pteron*: fin or wing; *novus*: new; *angliae*: England).
VERNACULAR NAMES	The name "humpback" comes from the distinctive hump on its dorsal fin and perhaps from the way it hunches up its back before showing its tail and diving. (Fr. Baleine à bosse, rorqual à bosse, Mégaptère and Jubarte.)

DESCRIPTION
Length: On average, 12 to 13 m (39 to 43 ft), maximum 16 m (53 ft). Some have reported specimens of more than 19 m (62 ft) long. The largest reported Humpback was a 26.8 m (88 ft) female that was caught in the Caribbean.
Weight: 25 to 30 tons

Color:	Its dorsal surface is dark grey or black; ventral surface is white with spots of black. The underside of the caudal fin is characterized by color patterns that are unique in each individual. This trait is important in photo-identification projects used in much the same as we use fingerprints to identify human beings.
Dorsal fin:	Small (not more than 30 cm or 12 in.), triangular or falciform and mounted on a hump. It is set about two-thirds of the way back from the tip of the snout.
Pectoral fin:	The very long flippers (up to one-third of the body length (i.e., 4 to 5 m or 13 to 16 ft) are white with black patches.
Head:	Small, but broad and flattened. It is characterized by an outgrowth on its chin called the jaw plate that is often encrusted with barnacles. Numerous knobs called tubercles (or sensory nodules) are found on its upper and lower jaws. Tubercle pattern varies among individuals but, in general, they seem to form three rows along the upper and lower jaws: one medially and two on both sides of the lower jaw. Each sensory nodule contains a vibrissa.
Baleen:	From 270 to 400 plates on each side of the mouth, the longest ones measuring between 85 and 104 cm (2.8 and 3.4 ft) in length and 30 cm (1 ft) wide. The bristles appear somewhat coarser than those of the Fin whale. It is dark grey or olive brown in color and the end-bristles are generally lighter.
Ventral grooves:	12 to 30 grooves on the throat and chest that extend to the navel

NATURAL HISTORY

Breathing sequence:	2 to 8 breaths at 20- to 30-second intervals
Dives:	Variable, from 3 to 28 minutes. It hunches up its back on its terminal dive (also called the 'humpbacked' dive) and shows its tail.
Blow:	Up to 3 m (10 ft). In the absence of wind, the spout is very recognizable because it is almost as high as it is wide. Let us keep in mind that the height of the blow depends on the level of the animal activity: high and dense when it is active and much more discrete when the animal is at rest, or performing short and shallow dives.

Diet: Herring, sandlance, capelin, mackerel, krill, and other planktonic crustaceans. These whales are famous for concentrating their prey in one area using the bubble-net feeding technique. While spiraling to the surface, they let out air bubbles that slowly rise to the surface creating a net of bubbles; they then project themselves vertically into the center of the net with their mouth wide open and capture the concentrated prey items.

Vocalizations: Produces a vast array of sounds: groans, moans, snores, whos, yups, chirps, ees, oos, and various clicks. In this regard, Humpbacks are probably the most studied cetaceans, being the ones responsible for the whale songs that we enjoy so much. At their breeding grounds, these complex songs are continuously repeated according to identifiable patterns that vary within the different populations, social groups and sometimes individuals as well. These variations are called dialects and are useful in identifying individuals. Furthermore, these patterns evolve with time. There are many theories that try to explain the role of these songs, but most evidence suggests that its principal function is sexual —isolated males may sing to attract females. These sounds range from 40 Hz to 5kHz.

Reproduction: Breeding and calving occur from January to March in the tropical waters of both hemispheres. Gestation lasts about 11 months. Newborns measure 4 to 5 m (13 to 16 ft) and weigh one to two tons. The milk of Humpbacks is very rich in fat (up to 41%). The calf takes up at least 43 kg (95 lbs.) of milk daily, and is weaned five months later. By then, the calf is between 7.5 and 9m (24.6 and 29.5 ft). Females give birth every two or three years, sometimes twice in three years. The Humpback whale of the North Atlantic breed in the waters of the Silver Bank, near the Dominican Republic. Those of the North Pacific mostly breed in the Hawaiian regions. Sexual maturity is reached when the animal measures 11 to 12 m (36 to 39 ft) in length, near the fourth year of life, sometimes sooner. Courtship involves social games in which a group of males compete for a female. These games can degenerate into violent confrontations.

Longevity:	Probably close to or more than 100 years
Population:	One of the species that has suffered most from whaling. Some populations seem to be increasing while others seem rather constant. This species should be considered as vulnerable and should be protected. There are approximately 12,000 to 15,000 individuals worldwide, but only a few hundreds in the Atlantic.
Distribution:	Worldwide.
Parasites & Enemies:	This slow moving animal is an ideal target for all types of ectoparasites and commensal organisms: whale-lice, diatoms, barnacles. It carries up to one ton at certain times of the year. The Humpback is also host to a number of internal parasites. Studies have shown that it can accumulate large concentrations of DDT, PCBs and other organochlorines in its fat. This can bring about the contamination of the calves during nursing.
Behavior:	On their feeding grounds: solitary, small groups, or sometimes huge groups of over 200 individuals. On their breeding grounds: mother and calf accompanied by a male, known as an escort, apparently for protection. In spite of its relatively slow speed, this whale can be highly active, with behavior patterns similar to those of other whales, but performed on a more frequent basis.

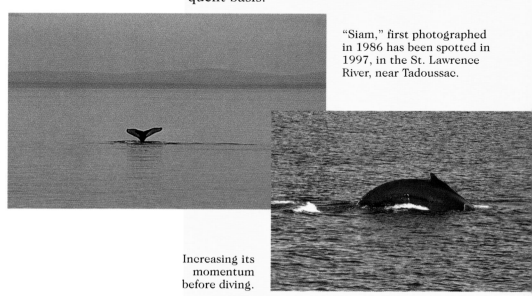

"Siam," first photographed in 1986 has been spotted in 1997, in the St. Lawrence River, near Tadoussac.

Increasing its momentum before diving.

Humpback breaching.   *Dimitri Marguérat*

Dimitri Marguérat

Fins of Humpback whales. By looking closely one can see barnacles and whale lice.

Dimitri Marguérat

ORES - Ursula Tscherter

325

The distinctive color patterns on the ventral side of the caudal fin of a few Humpback whales.

BREATHING SEQUENCE OF THE HUMPBACK WHALE

THE SEI WHALE

Balaenoptera borealis

ORDER	Cetacea
SUBORDER	Mysticeti
FAMILY	Balaenopteridae
SPECIES	*Balaenoptera borealis* (Latin: *borealis*: Northern).
VERNACULAR NAMES	Pollock whale, Coalfish whale, Sardine whale, Japan Finner, Rudolphi's Rorqual. The name "Sei" (Norwegian *seje*: coalfish, or pollock) was given because this whale had the habit of appearing at the same time as the coalfish. (Fr. Rorqual boréal, baleine du Nord.)

DESCRIPTION

Length:	12 to 16 m (39 to 52 ft). It is the second smallest of the rorquals (the Minke whale is the smallest).
Weight:	20 to 30 tons
Color:	The body is greyish-blue, dark grey or black, sometimes brownish, on its dorsal side and light grey underneath. Both sides of the lower jaw are of the same color, allowing us to distinguish it from the Fin whale.
Dorsal fin:	Similar to that of Minke whales, but is positioned a little more toward the front.
Head:	Bears only one central ridge, distinguishing it from the Bryde's whale which has three ridges (one central and two lateral).
Baleen:	From 318 to 340 plates in whales of the North Atlantic and 296 to 402 in those of the Southern Hemisphere.
Ventral grooves:	32 to 62. They do not reach the navel.

NATURAL HISTORY

Breathing sequence:	4 to 9 breaths at 40- to 60-second intervals, sometimes every 20 to 30 seconds for a period of 1 to 4 minutes.
Dives:	From 5 to 20 minutes. It does not hunch up its back before diving.
Blow:	The blow can be high.
Diet:	Krill, fish, and squid
Reproduction:	Breeding and calving occur during the winter months in tropical waters. Gestation lasts about 11 months. Newborns measure 4 to 5 m (13 to 16 ft) and weigh 700 to 800 kg (1,540 to 1,764 lbs.). The calf is weaned six to seven months later as it arrives with its mother at the feeding grounds. By that time, it measures 8 to 9 m (26 to 30 ft). Females can copulate again the following winter, so calving can take place every two years.
Longevity:	Examination of the earplugs suggests that they can live up to 60 years (Lockyer, 1974).
Population:	This species has been heavily exploited by the whaling industry. Only a few thousand remain in the North Atlantic. With a worldwide population of 54,000, it is on the endangered list.
Distribution:	Migratory species travel from the cold waters of the north where they feed to the tropical and

Behavior:

Parasites & enemies:

warmer waters of their breeding grounds. They are found worldwide in all oceans. Solitary or sometimes in small groups of 2 to 30 individuals. It rarely breaches, swims close to the surface, and is a rapid swimmer.
Whalers have often made mention of Killer whales attacking Sei whales. Being fast swimmers, they have very few external parasites. They are host to a number of internal parasites however, acanthocephalan, cestodes, nematodes, flukes, etc.

Near Percé, June 1997.

Élie Forté.

Williamson

THE TROPICAL OR BRYDE'S WHALE

Balaenoptera edeni

ORDER	Cetacea
SUBORDER	Mysticeti
FAMILY	Balaenopteridae
SPECIES	*Balaenoptera edeni*
	First described as a new species by Anderson in 1878.
VERNACULAR NAMES	Tropical whale, Eden's rorqual. (From Bryde, in honor of Johan Bryde a Norwegian who helped build the first two whaling stations in South Africa. From Eden, in honor of Dr. Ashley Eden who showed the first specimen to Anderson, who later identified it as a new species.) It was also named the Tropical whale because it was generally sighted in the more tropical and temperate waters. (Fr. Rorqual de Bryde, rorqual tropical, rorqual d'edeni)

DESCRIPTION

Length: On average, 14 m (46 ft). It is a relatively small whale often mistaken for a Sei whale, a Minke whale and even a small Fin whale.

Weight: More than 30 tons

Color: Both sides of it lower jaw are light-colored. It also has three prominent longitudinal head ridges, on either side of the usual central ridge. Its back is bluish black. Its belly is whitish. The color and pattern vary considerably from one individual to the next. The body often bears white patches that could very well be scars.

Dorsal fin: Falciform and is set about two-thirds of the way back from the tip of the rostrum.

Baleen: Dark grey, the first ones are rather whitish in appearance. From 250 to 365 plates.

Ventral grooves: 40 to 70 grooves that extend to the navel

NATURAL HISTORY

Breathing sequence: Irregular. Takes one to two breaths followed by a shallow dive, or 4 to 5 breaths leading to a deeper and longer dive.

Blow: The blow can be high.

Diet: Pelagic crustaceans, fish such as herring, anchovies, mackerel, and cephalopods.

Vocalizations: Powerful low-frequency moans (124 to 250 Hz).

Reproduction: Gestation lasts one year followed by a nursing period of six months. After a six-month rest, the female is ready to breed again. Calving can then take place every two years. Sexual maturity is reached between the ninth or thirteenth year of life. According to various authors, breeding occurs during the autumn months, while others believe it can take place anytime during the year. It seems likely that the populations that spend their whole life in tropical waters can breed year-round. Newborns measure 3.5 m (11.5 ft) and weigh 900 kg (1,984 lbs.).

Longevity: May live up to 50 years.
Population: Poorly known. Some people have suggested 90,000 worldwide.
Distribution: Between 40 degrees latitude north and 40 degrees latitude south. The range may be larger where waters reach temperatures higher than 15-20°C (59-68°F).
Behavior: Groups of more than four or five individuals are seldom seen. It can perform repeated breaches. A rapid swimmer, although a little slower than the Sei whale.

Fact sheets

THE CETACEA

THE ODONTOCETES

The Odontocetes

Odontocetes are toothed whales.

From the Greek: *odous* **meaning tooth.**

Toothed whales have anywhere from 1 to 250 teeth that are generally not differentiated into incisors, canines and molars, but are similar in appearance (homodont). They are sharp and used only to grasp prey. In some cases, these teeth seem to have one function only, that of establishing social hierarchy; teeth erupt only in males and are used during confrontations (more or less ritualized) with rival males.

Unlike Mysticetes, they have but one external nostril forming a single blowhole

6 families containing 66 species are found:
- *Monodontidae:* the Narwhal and the Beluga whale (white whale);
- *Phocoenidae:* the Harbour porpoise;
- *Delphinidae:* Pilot whales, dolphins, and the Killer whale;
- *Physeteridae:* the Sperm whale;
- *Ziphiidae:* Beaked whales;
- *Platanistidae:* River dolphins.

THE BELUGA WHALE OR WHITE WHALE

Delphinapterus leucas

ORDER	Cetacea
SUBORDER	Odontoceti
FAMILY	Monodontidae
SPECIES	*Delphinapterus leucas* Latin: *Delphinus*: dolphin; *pteron*: fin or wing; *a*: lack of, in reference to its not having a dorsal fin.
VERNACULAR NAMES	Beluga, White whale, Sea canary. (Fr. Béluga, marsouin blanc, marsouin (Québec.)

DESCRIPTION
Length:

The largest males can reach 6 m (20 ft). The largest specimens are found near Greenland or the Sea of Okhotsk. The St. Lawrence River is host to a small population, with an average length of 4.5 m (15 ft). The smallest specimens are found in the White Sea and Hudson Bay.

Weight:	Maximum 1,300 kg (2,866 lbs.), and on average 650 kg (1,443 lbs.). Forty-three percent (43%) of that weight is made up of skin, blubber and fins. Skin: The Beluga whale, along with the Narwhal, are the only cetaceans to have skin that can be tanned for various uses. Its skin is sturdy enough for tanning probably because it lives in ice-covered waters that it has to break through in order to breathe. As early as the seventeenth century, the French explorer Samuel de Champlain was aware of the quality of Beluga skin, robust enough to have been used, in those days, as a type of bulletproof jacket!
Color:	The adult is entirely white. Newborns are brown (café au lait). As they get older, their skin lightens through stages of blue or greyish-brown to become white around the age of 4 to 6.
Head:	Relatively small, but appears bigger because of the melon, the fatty pad of tissue on its forehead. The head is highly mobile. Contrary to other Odontocetes, it has a well-defined and flexible neck because its cervical vertebrae are not fused together. The skin of the head is especially thick, allowing the animal to break up to 10 cm (4 in.) of ice.
Teeth:	8 to 11 per 1/2 jaw (32 to 44 in all). These peg-like teeth rapidly wear down to the level of the gum.

NATURAL HISTORY

Breathing sequence:	Breathes 5 or 6 times at the surface
Dive:	Usually 1 minute but can last between 10 and 15 minutes for dives to depths of 647 m (2,123 ft; Ridgway *et al.*, 1984).
Blow:	Hard to detect but can be seen in very calm weather.
Diet:	Feeds on a wide range of organisms: fish (up to 25 kg (55 lbs.) of capelin daily, in the St. Lawrence), crustaceans, cephalopods, benthic organisms such as polychaete worms, for example.
Vocalizations:	Often called the sea canary because of its wide sound repertoire (whistles, chirps, moos, clicks, high-pitched trills, squeals, etc.). We can hear them from the surface of the water. They use their biosonar to locate holes in the ice.

Reproduction:	Sexual maturity: 5 to 8 years of age. Gestation: 14 months. Calving occurs during summer. Nursing lasts until the calf is two years old. The female nurses its calf during the initial stages of the subsequent gestation period. Females give birth every 36 months. Weight and length of calves at birth: 80 kg (176 lbs.) and 1.5 to 1.6 m (5.2 ft).
Longevity:	30 to 35 years
Distribution:	They are found in the Arctic and subarctic regions. The worldwide population is hovering at around 60,000 individuals, half of which are located in North American waters. The largest herd is found west of Hudson Bay, in the strait of Lancaster, and in the Beaufort Sea. The mouth of the Saguenay River being a subarctic enclave, a population resides there throughout the year and represents the southernmost limit to this species' distribution in North America. This herd's numbers have dwindled from 5,000 to about 500, following uncontrolled exploitation and the effects of pollution. This population is a relic of the population that must have occupied the Champlain Sea following the last glacial period. Most of the time, it inhabits the middle estuary, but in winter it moves eastward to Pointe-des-Monts (at the western limit of the Gulf of St. Lawrence). In the spring, summer and fall, these southernmost Belugas are found mainly from île aux Coudres and the Kamouraska region, in the west, to as far as Les Escoumins and île aux Basques further east. Most live in the vicinity of the mouth of the Saguenay River, from île aux Lièvres to Les Escoumins. The cold temperature of the water and the high productivity of this habitat, a result of current upwelling, explain the presence of Belugas in that region. During the winter months, they gather in the Pointe-des-Monts region on the North Shore and Sainte-Anne-des-Monts on the South Shore. Previously, their distribution in the St. Lawrence may have extended eastward to the Mingan Islands and Natashquan (Lower North Shore, Québec). The population of the St. Lawrence River is considered endangered, even though it has apparently ceased to dimin-

Behavior: ish. Necropsies performed on carcasses have revealed high levels of various contaminants such as organochlorines, PAHs, and heavy metals, affecting their immune system and make them susceptible to a plethora of opportunistic infections with often lethal repercussions.
They live in large herds subdivided into smaller groups, families composed of mothers and their young as well as larger groups (up to 500 individuals) of adult males.

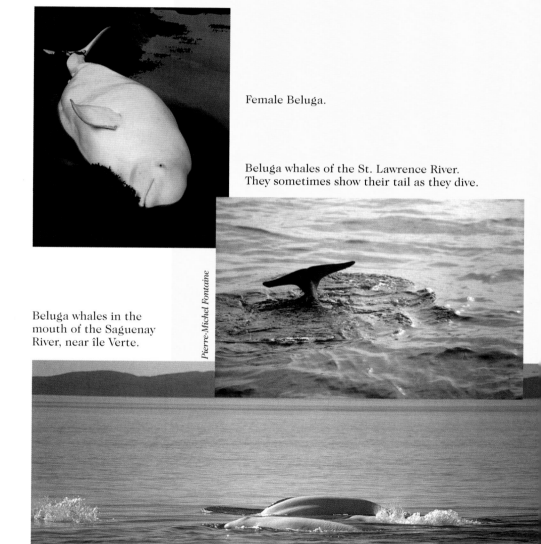

Female Beluga.

Beluga whales of the St. Lawrence River. They sometimes show their tail as they dive.

Beluga whales in the mouth of the Saguenay River, near île Verte.

Pierre-Michel Fontaine

339

Beluga under water. The blow hole is closed.

When at the surface it opens.

Sometimes the Beluga starts blowing when still under the surface of the water.

ORES - Ursula Tscherter

BREATHING SEQUENCE OF THE BELUGA WHALE

THE NARWHAL

Monodon monoceros

ORDER Cetacea
SUBORDER Odontoce*ti*
FAMILY Monodontidae
SPECIES *Monodon monoceros*
Greek: *mono*: one, *odous*: tooth, *ceros*: horn
VERNACULAR NAMES Sea-unicorn. (Fr. Narval, licorne de mer).

The second tooth is hidden in the maxillary.

DESCRIPTION

Length:	3.8 to 5 m (12.5 to 16.4 ft), excluding the tusk, which can measure up to 3 m (9.84 ft)
Weight:	800 to 1600 kg (1,764 to 3,528 lbs.)
Color:	The adult's back is greyish-white with back patches and its belly is white. Newborns are uniformly greyish-blue and lighten as they get older.
Head:	Small, with a pronounced melon. Males have a visible tooth (tusk).
Teeth:	The fetus has six pairs of dental papillae in the upper jaw and two pairs in the lower jaw. Only the first two in the upper jaw develop and persist throughout the animal's life. Generally, only one of these will erupt, in males, to reach astonishing sizes. We have seen some reach 3 m (9.8 ft) in length and weigh more than 10 kg (22 lbs.). The second tooth in males and both teeth in females remain in the maxillary, the pulp cavity closing itself off relatively early following birth. These teeth can reach 29 cm (11 in.) in length in males and 23 cm (9 in.) for the left (always the longest) in females. Some males, and some females for that matter, occasionally have two tusks. The function of the tusks is poorly understood. They are evidently not necessary for feeding since females (and sometimes males) lack tusks. Furthermore, they are extremely fragile and are therefore not very useful as a weapon. Since only males have tusks, we can conclude that they are used in establishing dominance. Their fragility could lead us to believe that confrontations are highly ritualized. However, the tusks are often broken and males bear many scars on the head, due perhaps to violent confrontations. There is mention of a broken tusk that had in its pulp cavity the tip of another tusk! In order for that to happen, the two animals would have had to swim toward each other at a considerable speed!

The tip of the tusk, whether broken or not, is always worn. This is not necessarily the result of any specific use. The tusk points downward and we know that Narwhals feed on benthic organisms. It is therefore normal that the tip of the tusk scrapes along the ocean floor as the animal chases its prey.

Skin: Very thick, particularly on the head allowing the animal to break the ice in order to breathe at the surface.

Skeleton: Similar to their cousins, the Belugas. Its cervical vertebrae are not fused together, allowing a very flexible neck.

NATURAL HISTORY

Breathing Sequence: Can breathe every 100 m (328 ft; 15 minutes) when migrating, but, in general, they breathe every 10 to 15 seconds and dive for periods of 7 to 8 minutes. Dives of 20 minutes are also possible. They can also perform dives of more than 350 m (1,148 ft).

Diet: Fish such as Arctic and Atlantic cod, Greenland halibut, squid, and shrimps.

Vocalizations: A gregarious species, it has a vast repertoire. Some sounds have social functions while others are used in echolocation.

Reproduction: Breeding occurs from March to May, and especially near the end of April. Gestation is about 14 months. Calving takes place in July-August. Nursing lasts 1 to 2 years. A female gives birth once every three years or so.

Longevity: Probably to the age of 50

Population: Between 25,000 and 40,000 worldwide

Distribution: Circumpolar species. Mostly from the north of Hudson Bay, in Davis Strait, Baffin Bay, and the Barents, White, de Kara and de Laptev Seas toward the east.

Behavior: Live in groups of 3-4, sometimes up to 10, individuals. There is a certain sexual segregation within the groups; males (bachelors) gather together. It is a migratory species that swims close to the shore. In Canadian waters, they winter in Davis Strait, then to the fjords of Baffin Island, to the west of Greenland and further north.

Parasites & enemies: Whale-lice and nematodes. Polar bears, Killer whales, and possibly some sharks are predators. Humans continue to hunt this species for food, but especially for its expensive tusks. Aboriginal peoples still hunt these whales, often with rifles from the shore. Therefore, only those that float or are killed in shallow waters can be salvaged. We may well ask if government quotas effectively protect this species from man's overexploitation, since those whales that cannot be salvaged are not included in the quotas.

Tooth of a male Narwhal. Three meters long, it is missing one meter broken off while the animal was living.

Narwhal teeth normally hidden in the premaxillaries of the male or the female. They normally never erupt.

Robert Michaud

345

THE SPERM WHALE

Physeter macrocephalus

ORDER	Cetacea
SUBORDER	Odontoceti
FAMILY	Physeteridae
SPECIES	*Physeter macrocephalus*
	Physeter: blower; *Macrocephalus*: big head
VERNACULAR NAMES	Sperm whale. The name is a reference to the clear liquid in the animal's head initially believed to be seminal fluid. This liquid was thus called spermaceti (etymologically: sperm of a whale). Could the animal's sexual prowess and the size of the male genitalia have led the first whalers to believe that the animal needed a huge sperm reservoir in order to reproduce, hence the name Sperm whale? (Fr. Cachalot.)

DESCRIPTION

Length: Males: on average, 15 m (49 ft); maximum 18.3 m (60.1 ft) females: on average, 11 m (36 ft); maximum 12.5 m (41.0 ft)

Weight: Heaviest female on record: 11 m (36 ft) weighing 24 t. The heaviest male was 18.1 m (59.4 ft) and weighed 57.1 t.

Color: Dark grey or brownish-grey. Some white or pale patches may be present on the upper lip, on both sides of the mandible, and on the belly above the genital area. The patches may increase in size the older the animal gets. Some specimens are completely white; they may be either albino or very old.

Head: Huge and blunt, more than one-third the total length of the animal. A single blowhole is set far forward, above the lower jaw, on the left. Even when seen from the front, the whale's hydrodynamic shape is evident. The eyes are small and located midway between the back and belly. The head often bears sucker marks left by the tentacles of the giant squid (*Architeutus sp.*) on which this whale feeds. Some observers have reported the existence of scars that measured 20 cm (7.9 in.) in diameter, which would indicate (unless they had grown along with the animal), in all likelihood, that the squid responsible for these scars was some 45 m (148 ft) in length!

Dorsal fin: Absent, replaced by a type of dorsal ridge, a series of bumps that decrease in size as they approach the tail.

Pectoral fins: Small, paddle-like and located close to the eyes

Keel: Behind the anus, there is a keel opposite to the dorsal ridge.

Skin: The skin is wrinkled.

Teeth: Two rows of 18 to 29 teeth protrude from the lower jaw. There are no teeth in the upper jaw, although there can be up to 11 pairs of rudimentary teeth in the upper gum. I noticed the presence of such teeth in a stranded juvenile male that measured 13 m (42.7 ft; île d'Anticosti, June 1992). These teeth did not penetrate the maxillaries but were visible nonetheless. Teeth probably erupt around the tenth year of life.

NATURAL HISTORY

Breathing sequence: The first blow, following a deep dive, is a virtual explosion and can be heard from a distance of almost a kilometre. The breathing sequence that follows can be described as a breath every 4 seconds: 3 to exhale and 1 to inhale. The Sperm whale stays at the surface, breathing 5 or 6 times per minute, and then dives.

Dives: According to Lyall Watson, ancient whalers estimated the size of the Sperm whale by correlating its respiratory cadence and dive time: a Sperm whale that breathed 50 times at the surface and then dived, staying underwater for 50 minutes, measured 50 feet. I had the opportunity see a Sperm whale that must have measured about 45 feet (just like the two specimens I examined on île d'Anticosti). Unfortunately, I could not establish a correlation between the number of breaths and its dive time.

Diet: Cephalopods, including the giant squid, and fish.

Ambergris: In the intestines of 5% of Sperm whales there is an impaction or concretion called ambergris. Its main constituent is an amberine ester ($C_{23}H_{40}O$). It is used in the perfume industry and was once worth more than its weight in gold.

Vocalizations: Stranded animals can produce a type of very powerful lowing sound. Underwater, they emit clicks composed of 1 to 9 pulses, all lasting from 0.1 to 2 ms. These clicks are emitted in series which can last up to 10 minutes without any changes in rhythm (generally 1 to 7 clicks per second).

The way that these pulses are ordered in a click is specific to each individual, constituting a sort of signature allowing for the identification of a whale. The sounds produced by Sperm whales are very likely used for echolocation and probably also have a social function, allowing individuals to recognize one another and to synchronize their activities.

Reproduction: Sexual maturity: Males: 18-21 years, 11 or 12 m (36.1 to 39.4 ft) in length. The combined testicular weight varies from 8 to 18 kg (18 to 40 lbs.; average 12 kg or 26 lbs.). The gestation period lasts 14 to 15 months (Rice, 1989). Nursing lasts from one year and a half to two years. Females calve every 4 to 6 years. Length and weight of newborn: 3.5 to 4.5 m (11.5 to 14.8 ft), 1 ton, respectively.

Social behavior: Sperm whales are distributed into two very distinct group types.

A. **Mixed groups or breeding groups (nursery school)** – They are made up of females of all ages and of males less than 20 years old that have not yet reached sexual maturity (sub-adult and immature males). Most of the time, the females are related and form stable groups. These groups remain in tropical or subtropical waters (they rarely go above 40° latitude, north or south) throughout the year. They consist of 4 to 150 individuals. The larger groups are probably aggregations of the smaller ones. The average number of individuals within a group is around 25.

B. **Groups of bachelor males, rather homogenous with respect to age** – The youngest (less than 12 m (39 ft) long) are composed of 20 to 30 individuals. The older, sexually mature males are more solitary or form small groups of up to 6 individuals. These animals travel to higher latitudes. The mature males join the breeding groups between January and August where the dominant bull breeds the available females. The contests to establish dominance within the groups can be brutal as the numerous scars on the older males would indicate.

Dives: Diving toward the bottom is done on average at a rate of 120 m/min (394 ft/min), but may reach up to 600 m/min (1,970 ft/min; Rice, 1989). Dives may last up to 90 minutes. One observer has reported that a group of 5 individuals stayed submerged a period of 138 minutes (Watkins *et al.*, 1985). According to Lockyer, more than 96% of the dives are less than 30 minutes. The deepest dive was estimated at 3,000 m (9,850 ft). This finding is based on the discovery of bottom-dwelling sharks, which lived at depths of 3,195 m (10,490 ft) in that

Parasites & enemies:

area, in the stomach of a Sperm whale. Another dive was recorded at a depth of 2,250 m (7,388 ft) using acoustic detectors.

The Sperm whale is host to many parasites: in the blubber (*Phyllobotrium delphini*), the intestines, and even the placenta, where we find *Placentonema gigantissima* that can measure, at the end of the gestation period, more than 1 cm (0.4 in.) in diameter, and 8.4 m (27.6 ft) in length, for males (females: 3.75 m; 12.3 ft). It is also host to diatoms, whale-lice, and barnacles. Lampreys and cookie-cutter sharks are "predators," and remoras occasionally hitch rides.

Sperm whales of all ages have very few real predators. Killer whales and the large epipelagic sharks are often seen on calving sites. The meat or blubber of Sperm whales found in the stomachs of these animals could also have come from the carcass of a dead animal.

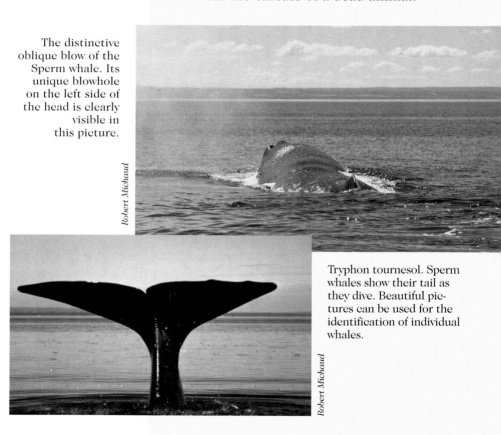

The distinctive oblique blow of the Sperm whale. Its unique blowhole on the left side of the head is clearly visible in this picture.

Tryphon tournesol. Sperm whales show their tail as they dive. Beautiful pictures can be used for the identification of individual whales.

The single blowhole is forward on the left.

The distinctive oblique blow of the Sperm whale. Its unique blowhole on the left side of the head is clearly visible in this picture.

In spite of its shape when seen from its side, the Sperm whale, seen from the front, has a truly hydrodynamic body.

Sperm whale, juvenile male, stranded in Anse à Valleau, Gaspesia, in September 2004. Its skeleton can be seen in the CIMM museum, Tadoussac

THE PYGMY SPERM WHALE

Kogia breviceps

ORDER	Cetacea
SUBORDER	Odontoceti
FAMILY	Physeteridae
SPECIES	*Kogia breviceps*
	Latin: short head
VERNACULAR NAMES	Fr. Cachalot pygmée, petit cachalot

DESCRIPTION
Length: 2.7 to 3.4 m (8.9 to 11 ft)
Weight: 400 kg (882 lbs.)
Color: Its back is bluish-grey while its sides are light grey. Its belly is whitish. Just before the pecto-

General body shape: ral fin there is a light crescent-shaped mark, which gives the impression of gills on the sides of the head.
A sturdy animal whose head is reminiscent of the profile of certain Catostomidae (carps, suckers). The mouth is located underneath the head. The dorsal fin is small, sickle-shaped and located just behind the midpoint of the body. The extremity of the caudal fin is concave and is slightly notched. The blowhole is on the left side of the tip of the snout as in all Physeteridae; only the left nostril is used in breathing.

Head: It is filled with spermaceti.

Teeth: There are 9 to 16 pairs of teeth in the thin and fragile mandible. There can be three pairs of rudimentary teeth in the upper jaw.

NATURAL HISTORY

We know relatively little about the two smaller sperm whales: the Pygmy Sperm whale (*Kogia breviceps*) and the Dwarf Sperm whale (*Kogia simus*). In fact, the acknowledgment of these two species' existence is very recent. The Pygmy Sperm whale is usually a solitary animal but may form small groups of 4 to 5 individuals. It is easily approachable when it is motionless at the surface, but seems fearful of boats and does not move too close to them. This slow-moving species lives close to the continental shelves, in the warm and temperate waters of the world. It is probably not as rare as field observations results would have us believe. When surprised, it secretes a reddish-brown liquid from its anus. This liquid probably serves as a cover or a lure, protecting it from its enemies, a technique that is similar to the use of ink by cephalopods. The lower part of its intestines can contain more than 12 litres (2.6 gal.) of this liquid.

Breathing sequence: Breathes once and then lets itself sink. This animal breaches.

Diet: Squid, bottom fish, crabs. The position of its jaws probably allows it to feed on the epifauna (i.e., benthic organisms that live on the surface of the seabed).

Reproduction: Mature males have large testicles and can produce great quantities of seminal fluids. Testicles can reach 50 cm (1.6 ft) in length and 5 kg (11 lbs.), as one 3.05 m (10 ft) specimen indicates (data provided by Marineland in Florida). These dimensions may include the epididymis. Gestation lasts 9 months (11 months, according to some authors). Breeding takes place in the summer while calving occurs in spring. The existence of lactating gravid females would indicate that these whales can calve every year. At birth, the calf measures 1.2 m (3.9 ft) and weighs 50 kg (110 lbs.). Nursing lasts about one year.

Population: Unknown

Vocalizations: The two Kogia species are not very vocal. They most likely emit clicks to locate their prey since they forage in the aphotic or dimly lit zones of the oceans.

CLOSELY RELATED SPECIES: *Kogia simus*

This species is closely related to *Kogia breviceps*, in fact, so much so that they had initially been grouped together as a single species. *Kogia simus* was only recently given status as a species (Handley, 1966). *Kogia simus* is smaller: 2.7 m (8.8 ft) in length and 272 kg (600 lbs.) in weight. Notable differences are seen in the skull (*Kogia simus*'s is similar to that of the Sperm whale, when seen from above) and in their dentition. *Kogia simus* has only 8 to 13 pairs of teeth in the lower jaw, and almost always 3 pairs in the upper jaw, whereas upper teeth are extremely rare in *Kogia breviceps*. *Kogia simus* appear to prefer warmer waters.

BEAKED WHALES
THE NORTHERN BOTTLENOSE WHALE

Hyperoodon ampullatus

ORDER	Cetacea
SUBORDER	Odontocetes
FAMILY	Ziphiidae
SPECIES	*Hyperoodon ampullatus*
VERNACULAR NAMES	In Greenlandic: *anarnak*: one that purges, in reference to the purgative quality of their oil. (Fr. Baleine à bec commune.)

DESCRIPTION

Length: Males 9 to 10 m (29.5 to 32.8 ft), females 7.5 to 8.5 m (24.6 to 27.9 ft)

Weight: Males 3.6 tons, females 3 tons

Color: The back is dark cinnamon-brown; the belly is slightly paler. Old males have a white patch on their head.

Dorsal fin: Falciform and set about two-thirds of the way back from the tip of the snout.

Caudal fin: Slightly concave with no visible notch

Pectoral fins: Small and pointed

Head: Pronounced sexual dimorphism: females have a relatively small melon, uniform in color; the bony ridges on their maxillaries are hardly developed. Males have a more distinctive melon, characterized by a white patch (whalers called them "white foreheads"), and have well-developed bony ridges on the upper jaw. This dimorphism is so pronounced that some authors, relying only on the available skulls, believed that males were of a different species (J. Gray, 1864), *Hyperoodon latifrons,* and even a distinct genus, *Lagenocetus.* In 1883, David Gray finally solved the problem by publishing a series of skull drawings of *Hyperoodon ampulatus* and the hypothetical *L. latifrons,* all belonging to *Hyperoodon ampulatus.*

Teeth: One pair of teeth at the tip of the lower jaw. They erupt only in males and are limited to a social function. Rudimentary teeth (about forty) can be found in the lower and upper jaws. I found seven pairs in the lower jaw of an immature male that washed ashore on the sand bars of Saint-Roch-des-Aulnaies (Québec), in November 1993 (Fontaine, 1996). I could not find any in the maxillaries, but this could have been a result of the conditions under which the dissection was completed.

NATURAL HISTORY

Breathing sequence: One blow every 30 or 40 seconds followed by a dive that lasts about 35 to 40 minutes, but that can last up to two hours. The animal shows its tail before performing deep dives. On such occasions, the spout may reach up to 2 m (6.6 ft) in height; it looks like a small, round and dense cloud.

Diet:	Squid (*Gonatus fabricii*), herring and benthic organisms.
Vocalizations:	Whistles, clicks, and chirps. Stranded, they emit moans and high-pitched sounds.
Reproduction:	Sexual maturity is reached toward the age of seven, when males measure 7.3 m (23.9 ft) and females 6,7 m (22 ft). Gestation lasts 12 months. Females have one calf every three years or so. At birth, the calf measures 3.5 m (11.5 ft) but its weight is unknown. Nursing lasts at least one year, perhaps longer.
Population:	Once very common, it has been excessively hunted and its numbers have decreased considerably. Little is known about this species' current status. It is considered a vulnerable species.
Behavior:	Inquisitive animals, Northern Bottlenoses have a tendency to approach boats. They will occasionally slap the surface of the water with their tail, and show their flukes prior to a deep dive. They live in mixed groups during the reproductive season; sexually mature males have a tendency to form homogeneous groups afterwards. They are highly protective and will stay with wounded companions. This particular characteristic made them easy targets for whalers during whaling.
Distribution:	Northern Atlantic, from Nova Scotia to the west coast of Spitzberg Island, near Greenland, in waters where depths reach 1,000 m (3,281 ft).

OTHER BEAKED WHALES

CUVIER'S BEAKED WHALE,
also known as Goose-Beaked Whale

Ziphius cavirostris

It can be confused with the Northern Bottlenose whale, due to the overlapping distribution range of the two species. Its forehead is not as pronounced and its beak is much shorter. Furthermore, there is an obvious indentation in the back of the blowhole. Males have two small but conspicuous teeth at the tip of their mandible.

As with all beaked whales, Cuvier's beaked whale is timid and poorly known. It feeds on squid and fish, frequents waters of depths that reach 1,000 m or more (3,281 ft), and consequently, it rarely approaches shore. This probably explains why these whales are seldom seen, although the stranding rates show that this species is probably on one the most abundant species of beaked whales in the world.

Length: 5 to 7 m (16.4 to 23 ft)
Weight: 2-3 tons

The V-shaped grooves, on the throat, is characteristic of this species.

The pectoral fin is relatively small.

The dorsal fin is curved.

The caudal fin is unnotched at its center.

A Northern Bottlenose whale. The author of this book is close to the animal and gives us an idea of the animal's size, November 1993.

At the tip of the right mandible, two teeth: the largest would have erupted if the animal had survived, the other one would never have developed since its root had already closed.

The vestigial maxillary teeth are discernible through the gum.

Vestigial teeth in the left maxillary. They were also between 33 and 46 cm (1.08 to 1.51 ft) from the tip of the maxillaries.

A young Hyperoodon ampullatus male—hipbones—the innominate bones.

The teeth from the mandible of a young male Northern Bottlenose whale (less than a year old) that beached with its mother, near Montmagny, November 1993.

Northern Bottlenose whales are teuthophagous (squid eaters). The stomach of this specimen was filled with many beaks of squids. The beaks are made of keratin and are completely indigestible, they are eventually evacuated with the feces, or regurgitated.

Young Northern Bottlenose whale that beached at Sept-Îles, September 10, 1997. As with all beached males, the penis has slipped from the genital slit. This whale was taken for a necropsy at the Maurice Lamontagne Institute (Oceanographic Research Centre, Sainte-Flavie). The skeleton was salvaged.

September 1997, measurements taken before performing necropsy on this young Northern Bottlenose whale.

THE HARBOUR PORPOISE

Phocoena phocoena

ORDER	Cetacea
SUBORDER	Odontocetes
FAMILY	Phocoenidae
SPECIES	*Phocoena phocoena*
VERNACULAR NAMES	Common porpoise. The Romans called it *"porcus piscus"* or pigfish. (Fr. Marsouin commun, pourcil.)

DESCRIPTION

Length:	Maximum: 1,8 m (5.9 ft)
Weight:	27 to 88 kg (59.5 to 194 lbs.)
Color:	The back is black, the sides are greyish and stomach is white.
Dorsal fin:	Triangular, set at the midpoint of the animal
Pectoral fin:	Small and oval
Head:	Small with no beak
Teeth:	22 to 28 pairs on the upper jaw, 21 to 26 pairs on the lower jaw. They are spatulate, not sharp.

NATURAL HISTORY

Breathing sequence: While hunting, 4 breaths at 10- or 20-second intervals followed by dives that generally last 2 to 6 minutes. While moving, they can breathe up to 8 times in a row, at one minute intervals. Since this is a small species and its dorsal fin is not very high, it is difficult to observe it when it comes to the surface to breathe, unless the water is extremely calm.

Diet: Feeds on a variety of prey: fish such as herring, mackerel, cephalopods (squid), and shrimp, up to 4 to 5 kg (8.8 to 11 lbs.) daily.

Vocalizations: Low-frequency sounds, clicks used in echolocation.

Reproduction: Sexual maturity is reached in four-year-old males, and in eleven-month-old females. Gestation lasts 11 months. Calving occurs from May to July, and usually one calf (75 cm or 2.5 ft) is born, although twins have been reported. Females will give birth once every two years. Nursing lasts eight months.

The Harbour porpoise has the highest testicular weight/body weight ratio of all cetaceans. (Barrette-P. M. Fontaine, 1993). A four-year-old male weighing 44.6 kg (98.3 lbs.), accidentally caught on August 23rd had a combined testicular weight of 3.202 kg (7 lbs.) or 7.2% of its body weight, almost as much as that of a fifty-ton Fin whale! An adult male in the peak of its breeding season has, on average, a combined testicular weight exceeding 2 kg (4.4 lbs.). Testicular size decreases considerably between rutting (breeding) seasons. Why such large testicles? To allow ejaculation of huge quantities of spermatozoa, in a series broken by very brief intervals. In mammals, males tend to limit their involvement in propagation to breeding, leaving postnatal care to females. To ensure their reproductive success, males therefore have to compete among themselves for the available females. Dominance over other males is achieved by becoming very large and strong, by developing horns or conspicuous attributes, or by manufacturing the largest possible amount of spermatozoa. This last method allows the male Harbour porpoise to copulate many times with the same female, or with many different females, in a relatively short period of time. The high amount of ejaculated spermatozoa, each time, reduces (to a minimum) the chances of successful fertilization by rivals that would later copulate with the same female. There is no sexual dimorphism in this porpoise (or if there is one, it most likely favors the female); males are not endowed with special attributes or colors. It is therefore likely that their reproductive success is not ensured by a competition that precedes breeding, but by producing the largest quantities of spermatozoa. This probably explains the enormous size of their testicles (and their accessory organs: epididymis and prostate as well).

Longevity: Probably 12-13 years

Distribution: Temperate coastal waters of the Northern Hemisphere

Population: Unknown. It is a species difficult to observe. It is probably also declining, being vulnerable for many reasons. Its proximity to the coast and the fact that it feeds on fish positioned high in the food chain bring about a higher level of contamination from various toxic substances. It is also very often accidentally caught in fishing nets.

Behavior: Harbour porpoises are timid; they avoid boats and do not ride the bow wave. These characteristics combined with their small size make them difficult to watch, except on very calm seas. Singles, pairs or groups of 5-6 individuals are usually seen. They exhibit strong social ties. They rarely jump out of the water. Their coastal distribution makes it a species that often beaches.

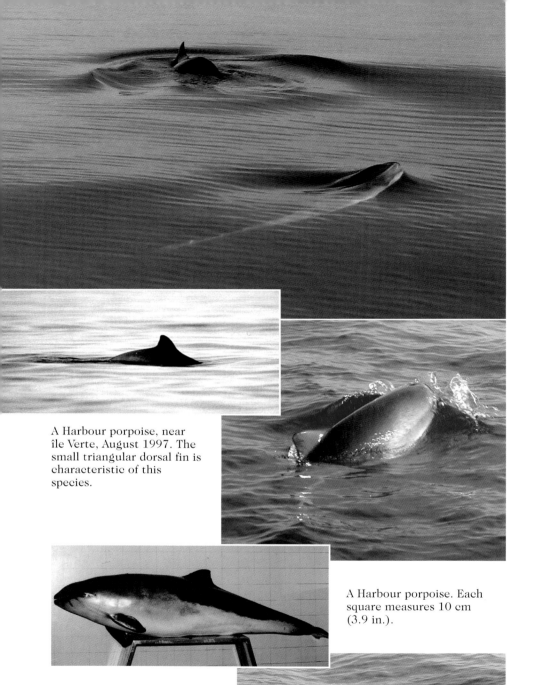

A Harbour porpoise, near île Verte, August 1997. The small triangular dorsal fin is characteristic of this species.

A Harbour porpoise. Each square measures 10 cm (3.9 in.).

A Harbour porpoise, near Percé, Gaspé Peninsula. Its small size makes observing this animal very difficult.

THE PILOT WHALE OR LONG-FINNED PILOT WHALE

Globicephala melaena

ORDER	Cetacea
SUBORDER	Odontoceti
FAMILY	Delphinidae
SPECIES	*Globicephala melaena* Latin: *globus*, balloon; Greek: *kephale*, head; *melanos*, black.
VERNACULAR NAMES	Long-finned Pilot whale, blackfish, and pot-headed whale. (Fr. globicéphale noir, baleine pilote.)

DESCRIPTION
Length: On average, males reach 6 m (19.7 ft) and a maximum of 8 m (26.2 ft). Females reach lengths of 4 to 6 m (13.1 to 19.7 ft)
Weight: 1.800 to 3.500 kg (3.968 to 7.716 lbs.)
Color: All black, sometimes with grey bands near the eyes. It has a distinguishing greyish-white, anchor-shaped mark on its throat.

Dorsal fin:	Set forward (in the front half of the back), low but wide at the base. In females, it is more triangular, whereas males have a more sickle-shaped dorsal fin.
Pectoral fins:	Very long (18 to 27% of the total body length), falciform and sharp flippers. The elbow joint is undetectable on the leading edge.
Head:	Bulbous melon (from which its name comes). Short snout
Teeth:	8 to 13 pairs in each jaw. They are located in the front of both jaws, are slightly curved, sharp and fit in between each other perfectly.

NATURAL HISTORY

Breathing sequence:	A few quick breaths before diving for a period ranging from 5 to 10 minutes
Dives:	At least to 600 m (1,968 ft)
Diet:	Almost exclusively teuthophagous, but also feeds on fish such as cod, herring, mackerel, etc. It eats up to 14 kg (31 lbs.) on each of its three daily feeding sessions.
Vocalizations:	Pilot whales are gregarious, living in family units, and have a wide vocal repertoire, including shrills, whistles, chirps, buzzing sounds, and snores complete its repertoire (social functions). Clicks are used for echolocation.
Longevity:	Age can be accurately determined since it was established that two dentine layers are added each year on their teeth. We now know that they can live up to 50 years.
Reproduction:	Females reach sexual maturity at the age of six, whereas males become sexually active toward the age of twelve. Gestation lasts about 16 months. Calving occurs between July and October, every three years. Nursing lasts 20 months. The calf begins ingesting solid food toward the tenth month of life.
Distribution:	A population in the northern Atlantic, to 38 degrees latitude north, and a population in the Atlantic and southern Pacific associated with the Humboldt (Peru), Falkland, and Benguela Currents. These two populations are separated by a wide band of equatorial and tropical waters.

Population: Unknown. Pilot whales have been excessively hunted ever since whalers first noticed how easy it was to frighten and force them into a group, and then to drive them toward the shore. Mitchell has estimated that 50,000 were killed in Newfoundland from 1951 and 1961 (10,000 in 1956 alone). Only one day was needed to catch 1,500 in Cape Cod, November 17, 1884. The population is probably rising now that whaling is only practised only in the Faeroes (around 1,700 killed yearly), as the rising number of strandings would indicate.

Behavior: They live in family units called pods containing more than 100 individuals. These very tight-knit pods are vulnerable to natural mass stranding, and to co-ordinated herding by whalers, done in order to drive them toward the shore or into sandy bays. DNA studies on complete herds driven ashore in Faeroese traditional whaling have permitted a better understanding of the social structure of these animals (Amos *et al*, 1993). They reveal that a pod is made up of related individuals that tend to remain together; males and females stay with their mother on a permanent basis. Furthermore, these studies have shown that the males within a pod never breed with the females within that same group. Breeding must take place when different pods gather together; males of one group join the available females of another pod. Groupings of hundreds, even thousands of specimens have been observed during certain periods of the year. This behavior is unusual. In gregarious animals, sexuality mature males normally ensure their reproductive success by competing for access to the females. Harems are formed where the dominant male chases away other males or prevents them from breeding (as is the case for wolves, for example). We do not really understand the advantages for male Pilor whales to behave in an opposite fashion. Perhaps, since access to females of other groups is not limited, it is more advantageous to help the highest number of known parents rather than to invest energy in protecting their immediate descendants.

They often spy-hop and tail-slap but seldom breach.

The pod's cohesiveness is responsible for the mass strandings that are often referred to as collective "suicides" (see the chapter on strandings).

Pilot whales have a hook-shaped dorsal fin.

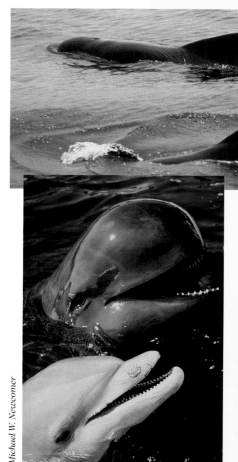

The melon of Pilot whales is responsible for the name "pot head."

Stomach contents of a Pilot whale stranded on Cotentin (Normandy, France) shore.

A Pilot whale at an aquarium in Florida.

Pilot whales are a gregarious species known for their mass strandings.

The skeleton of a Pilor whale in St. John's, Newfoundland.

THE SHORT-FINNED PILOT WHALE

Globicephala macrorhynchus

VERNACULAR NAMES	Blackfish. Fr. globicéphale tropical. It is an animal that is very similar to its cousin, the long-finned Pilot whale. Where they are seen together, it is almost impossible to tell them apart. As its name suggest, its pectoral fins are shorter, representing only 14 to 19% of the total body length. This species behaves very much like *Globicephala melaena*. (Fr. globicéphale tropical.)
Distribution:	Found in tropical and subtropical waters, between 40 degrees latitude, both north and south, throughout the world.
Skull:	One effective way to distinguish these Pilot whales consists in examining the manner in which the upper and lower jaws are positioned with respect to one another: if the premaxillaries (or premaxillae) completely cover the maxillaries (or maxillae), we are dealing with *Globicephala macrorhynchus*. On the other hand, if the maxillaries are clearly visible underneath and on the sides, then it is a *Globicephala melaena*.

Short-finned Pilot whale, Mauritania 1995.

THE KILLER WHALE

Orcinus orca

ORDER	Cetacea
SUBORDER	Odontoceti
FAMILY	Delphinidae
SPECIES	*Orcinus orca*
	Latin *Orcus*: Another name for Pluto, god of the underworld, also called Hades, a sort of demon.
VERNACULAR NAMES	Orca. In Norwegian. It is called *spekkhogger* for "blubber-chopper," in reference to the fact that these whales ate the blubber of dead whales that whalers left floating (once filled with air) before being hauled aboard the floating factories. (Fr. Orque Épaulard, baleine tueuse, orca, Killer whale.)

DESCRIPTION

Length: Females rarely exceed 7 m (23 ft), whereas males can reach up to 9.8 m (32 ft) in length.

Weight: From 4 to 8 tons

Color: The back is black, with a white patch above and behind the eye and a grey saddle behind the dorsal fin. Its belly extending up to the flanks is white. The white and black zones are clearly demarcated.

Dorsal fin: Triangular and particularly tall, especially in males. It is set midway along the body. Both sides are practically equal in males, whereas in females, the leading edge is curved while the trailing edge is almost vertical or slightly concave.

Pectoral fins: Large and paddle-shaped (1/5 the body length in old males)

Caudal fin: White underneath

NATURAL HISTORY

Breathing sequence: A few breaths at the surface generally followed by a shallow dive lasting anywhere from 3 to 5 minutes.

Diet: Killer whales are super predators, positioned at the top of the food chain. They feed on a variety of prey: marine mammals (true seals, eared seals and cetaceans), seabirds, fish (salmon, herring, etc.), and squid. In spite of their reputation for being ferocious animals, no documented cases of attacks on humans exist.

Vocalizations: Since Killer whales are gregarious, they have at their disposal a wide range of whistles, squeals, squeaks, etc., used in maintaining pod cohesiveness. These whales also use clicks for echolocation.

Each individual within a pod emits recognizable sounds, as a kind of acoustic signature. These individual sounds can be specific to one pod, which, by the way, can also have its own distinct signature. Using a hydrophone, a trained observer would then be able to identify the whales making these sounds without having to see them.

Reproduction: Breeding takes place from December to February. Sexual maturity is attained at 6 m (20 ft) for males, at 5 m (16 ft) for females. Gestation lasts at least one year, perhaps up to 16 months. Females give birth to one calf every three years or so.

Longevity: Male Killer whales may live up to 50 years, while the female may live 80 years or longer.

Population: Perhaps 50,000 worldwide. It is no longer hunted, but can be killed locally by fishermen who see them as possible rivals.

Distribution: Probably the most ubiquitous of cetaceans, it is found throughout all the oceans of both hemispheres, from the equator to ice floes.

Behavior: It is a gregarious species, living in stable and mixed pods of 5 to 20 individuals. A female generally serves as leader of the group. It is composed of 20% adult males, 20% sub-adults, with the rest being made up of young females and males. There is a strong cohesiveness within the group and cooperation among individuals has often been observed. Killer whales are fast swimmers (up to 50 km/hr; 31 mi./hr) that frequently breach. They also slap the surface of the water with their tail.

A male Killer whale in the Victoria Aquarium (BC). The curved dorsal fin is common in captive Killer whales.

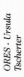

Female Killer whale, Tadoussac, summer 2003.

373

Female Killer whale, Tadoussac, 2003. They had not been seen in the area since 1967.

A pod of Killer whales, near Mingan. The dorsal fin of the male has been damaged.

Female Killer whale. The saddle, a white mark behind the dorsal fin, is different in all individuals and is used in identifying specific Killer whales.

Measurements were taken before salvaging the skeleton of a young Killer whale, Nouadhibou, Mauritania. To the right of the author, on the picture, we see Doctor Ba Abou Sidi and Mr. Sō of Nouadhibou's CNROP.

THE WHITE-SIDED DOLPHIN

Lagenorhynchus acutus

ORDER	Cetacea
SUBORDER	Odontoceti
FAMILY	Delphinidae
SPECIES	*Lagenorhynchus acutus*
VERNACULAR NAMES	Fr. Dauphin à flancs blancs, Lagénorhynque à flancs blancs.

DESCRIPTION
Length: 1.9 to 2.5 m (6.2 to 8.2 ft)
Weight: 150 to 200 kg (331 to 441 lbs.)
Color: The back is black, the belly white with pale grey bands on the flanks. A white band runs underneath the dorsal fin and a yellowish stripe is found on the caudal peduncle of adults.

Dorsal fin: Large and sickle-shaped. This particular shape is more pronounced in adult females.
Head: Small but pronounced beak. The upper part of the beak is dark while the lower one is white or light grey.
Teeth: Conical: 58 to 80 per jaw

NATURAL HISTORY

Breathing sequence: Breathes every 10 to 15 seconds, by skimming the surface or by breaching
Diet: Squid, fish, and sometimes benthic crustaceans.
Reproduction: Gestation lasts approximately 10 months; calving occurs during spring or at the start of summer. The calf measures from 1 to 1.3 m (3.3 to 4.3 ft) in length and weighs from 30 to 35 kg (66 to 77 lbs.).
Behavior: A gregarious animal, living in groups of 5 to 50 individuals. Herds of several hundreds, even thousands, have been observed. A certain degree of segregation seems to exist among the various groups. By analyzing the composition of stranded herds, researchers have noticed that some age groups were missing (2.5 to 6 years old), leading us to believe that sub-adults are excluded from the reproductive group only to join it later when sexual maturity is reached. This species is known to strand often.
Population: Unknown, but it does not seem threatened. This species has been rarely hunted, except in Norway.
Distribution: In the cold temperate waters of the northern Atlantic, from Greenland westward to the latitude of New York and eastward to the English Channel.

The White-sided dolphin is extremely acrobatic.

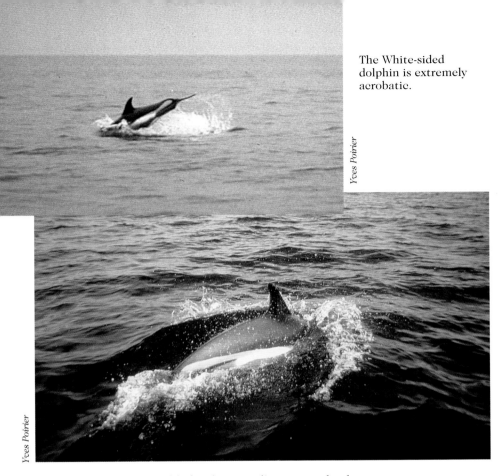

The white stripe responsible for the animal's name is clearly seen on this picture (île Verte, September 6, 1999).

THE WHITE-BEAKED DOLPHIN

Lagenorhynchus albirostris

ORDER	Cetacea
SUBORDER	Odontoceti
FAMILY	Delphinidae
SPECIES	*Lagenorhynchus albirostris*
	Greek: *lagenos*: bottle, *rhynchos*: beak
	Latin: *albus*: white, *rostrum*: snout
VERNACULAR NAMES	Fr. Dauphin à nez blanc, lagénorhynque à bec blanc.

DESCRIPTION
Length: 2.5 to 3 m (8.2 to 9.8 ft). This species is more robust, more massive than the White-sided dolphin.
Weight: 180 to 275 kg (397 to 606 lbs.)

Color:	The tip of its snout is white, the back is black and the belly is white with grey-white stripes on the sides. There is a grey-white patch on the upper part of the caudal peduncle, behind the dorsal fin.
Dorsal fin:	Large and falciform
Head:	Short beak, black on top
Teeth:	44 to 56 per jaw, conical

NATURAL HISTORY

Breathing sequence:	Breathes every 10 to 15 seconds, while staying close to the surface. May ride the bow waves of ships, breach and somersault. It is a fast swimmer.
Diet:	A variety of fish, squid and benthic crustaceans.
Reproduction:	Gestation lasts about ten months. Calving occurs during summer. Newborns measure from 1.2 to 1.6 m (3.9 to 5.2 ft) and weigh about 40 kg (88 lbs).
Distribution:	The cold temperate waters of the northern Atlantic, westward to the latitude of Cape Cod and eastward to Portugal.
Behavior:	This is a gregarious species, living in groups of 2 to 30 individuals, but herds of up to 1,500 individuals have been observed. It winters in the southern limit of its distribution range.

Jon Lien

THE COMMON DOLPHIN

Delphinus delphis

ORDER	Cetacea
SUBORDER	Odontoceti
FAMILY	Delphinidae
SPECIES	*Delphinus delphis*
VERNACULAR NAMES	Crisscross dolphin, saddle-backed dolphin. (Fr. Dauphin commun.)

DESCRIPTION
Length: 1.7 to 2.4 m (5.6 to 7.9 ft)
Weight: 70 to 110 kg (154 to 242 lbs.)
Color: Highly colored: the back is dark brown or black. This coloration forms a V underneath the dorsal fin. On the sides, an hourglass pattern

Dorsal fin:	Triangular to falciform, tall, and set midway along the animal.
Head:	A long and pronounced stripe between the forehead and beak characterizes this animal.
Teeth:	160 to 240 per jaw
Skeleton:	Two deep grooves in this animal's palate distinguish it from other species.

is seen, yellowish in front and light grey from behind. The underbelly is white. These visible colors aid group cohesiveness when they travel in high numbers.

NATURAL HISTORY

Breathing sequence:	Breathes every 10 to 15 seconds, then dives for 1 to 2 minutes.
Diet:	Small shoaling fish (anchovies, herring) and squid
Vocalizations:	A gregarious species. Pod cohesiveness is maintained through a variety of whistles and groans. Echolocation clicks are also used.
Reproduction:	Breeding takes place in spring; gestation lasts from 10-12 months. At birth, the calf measures 80-90 cm (2.6 to 2.9 ft).
Population:	It is probably one of the most abundant cetacean species. Some populations, however, like the one in the Black Sea, may be threatened by excessive hunting.
Distribution:	Worldwide in waters where temperatures do not fall below 10°C.
Behavior:	This animal lives in groups that sometimes include hundreds of thousands of individuals. One herd was seen to cover a surface area of 50 km (31 mi.) long and 800 m (0.5 mi.) wide (Robson 1976)! They are known to ride the bow waves of the fastest ships for hours. They are powerful swimmers and can reach speeds exceeding 22-25 knots (40.7-46.3 km/hr; 25.3-28.7 mi./hr).

A dolphin riding the bow wave of a boat.

THE STRIPED DOLPHIN

Stenella coeruleoalba

ORDER	Cetacea
SUBORDER	Odontoceti
FAMILY	Delphinidae
SPECIES	*Stenella coeruleoalba*
VERNACULAR NAMES	Blue-white dolphin, Euphrosyne dolphin. (Fr. Dauphin bleu et blanc, dauphin de Thétis.)

DESCRIPTION
Length: 1.8 to 2.5 m (5.9 to 8.2 ft)
Weight: 90 to 150 kg (198 to 331 lbs.)
Color: The back is dark brown becoming bluish following death. The flanks are light grey with a black stripe extending from around the eye all the way to the tail. It differs from the Common dolphin by not having an hourglass pattern on its flanks. The belly is white or pink.
Dorsal fin: Midway along the animal, tall, and falciform
Head: Forehead is not pronounced but a conspicuous furrow separates it from the well-defined beak.
Teeth: 78 to 110 per jaw, conical

NATURAL HISTORY
Breathing sequence: Breathes every 10 to 15 seconds, then dives to a few hundred metres for 5 to 10 minutes.
Diet: Small fish (herring, anchovies and mackerel) and squid

Reproduction: Gestation lasts 12 months, nursing 18 months even though the calf starts eating solid food at the age of 3 months. At birth, it measures 1 m (3.3 ft). Sexual maturity is reached at the age of 6.7 in males, and 8.7 in females.

Longevity: Perhaps 50 years (Kasuya and Miyazaki, 1975)

Population: Unknown, but it is one of the most abundant cetacean species living today. The fact that it is regularly caught in tuna hunts and that it is excessively hunted in Japan could make certain populations vulnerable.

Distribution: In the tropical, subtropical and temperate waters of the world. A beached specimen was found in the St. Lawrence river, giving weight to the authenticity of sightings all the way to Greenland.

Behavior: Gregarious animals, Striped dolphins can travel in groups of several thousand. In some areas, they will ride the bow waves of ships, while in others they seem to shun boats. They follow tuna in their movements, which often leads to their capture and death. They are easily driven to shore where they can be caught.

Steeve R. Baker

THE BOTTLENOSE DOLPHIN

Tursiops truncatus

ORDER	Cetacea
SUBORDER	Odontoceti
FAMILY	Delphinidae
SPECIES	*Tursiops truncatus*
VERNACULAR NAMES	Fr. Tursiops, tursion, grand dauphin, souffleur, dauphin à nez en bouteille.

DESCRIPTION:
Length: 1.9 to 3.9 m (6.2 to 12.8 ft)
Weight: 150 to 650 kg (331 to 1,433 lbs.)
Color: The back is dark grey or greyish-brown. The belly is lighter.
Dorsal fin: Set in the center of the back, tall, and sickle-shaped.

Head: Melon visible, pronounced furrow between the melon and the relatively long beak.

Teeth: 40 to 52 in the upper jaw, 36 to 48 in the lower jaw. The stout teeth are conical but wear down with age.

NATURAL HISTORY

Breathing sequence: Only 0.3 second is needed for the Bottlenose dolphin to exhale and inhale. It performs a series of respirations that are often accompanied by leaps before diving for 2 to 5 minutes. Tuffy, a dolphin at the Laboratory of Biosciences of the U.S. Navy at Point Mugu, in California, had been trained to photograph itself at ever-increasing depths. It took this animal 7 minutes and 15 seconds to reach a depth of 300 m (984 ft) and to come back up. This probably represents the longest time these dolphins can remain submerged.

Diet: Fish such as the mullet and other shoaling species, squid, and pelagic shrimp.

Vocalizations: The Bottlenose dolphin is a gregarious animal that has a vast repertoire of sounds and ultrasounds used in social interactions and in echolocation. It emits whistles, groans, squeaks and series of ultrasonic clicks. It is by examining the behaviour of Bottlenose dolphins in captivity that the notion of the use of sound to navigate has made its way into the scientific community. Since 1953, many scientists have worked on that theory, and significant progress, through the use of these animals, has been made.

Reproduction: Breeding takes place in the spring. Gestation lasts 12 months. At birth, the calf measures between 98 and 126 cm (3.2 and 4.1 ft) and weighs between 30 and 35 kg (66 and 77 lbs.). Nursing lasts 12 to 15 months even though the calf starts eating solid food well before that.

Longevity: 35 years or more.

Population: Unknown, but this species is very abundant.

Distribution: In the tropical, subtropical and temperate waters of the world. There are pelagic and mobile populations as well as more sedentary coastal ones.

Behavior: Bottlenose dolphins live in family groups that can come together to form herds of several hundred individuals. They take part in co-operative hunts for fish and show great adaptability. They rush toward shrimp boats when the fishermen pull out their nets, and salvage the non-commercial fish that have been thrown back into the sea. In some countries, Mauritania for example, they join with the Humpbacked dolphins (*Sousa teuszii*) to drive mullets into the fishing nets of local fishermen known as the Imragens, who hit the surface of the water with their sticks. Dolphins and fishermen both benefit from this type of activity. The Bottlenose dolphin is known to approach humans and play with them.

Fabrice Kerleau

Fabrice Kerleau

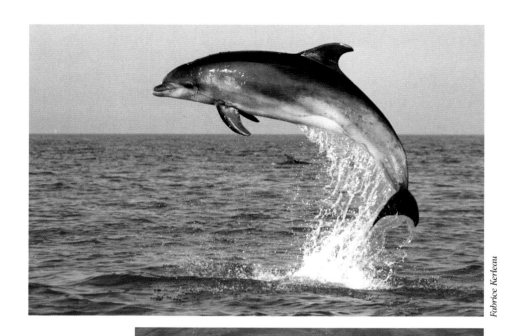

Bottlenose dolphins rapidly feel at home in aquariums.

Fabrice Kerleau

Fact sheets

PINNIPEDS

Pinnipeds

Order
As they are really unique compared to the other carnivores some taxonomists (Eisenburg, 1981; Novak and Paradiso, 1883) have placed them in a special order: Pinnipeds. Others (King, 1983) put them in the Order *Carnivora* and the Suborder *Pinnipeds*.

Super Family
They are divided into two super families *Otarioidae* and *Phocoidae*.

Family
Otarioidae have only one family: the *Otariidae*, and the *Phocoidae* are divided into two families *Phocidae* and *Odobenidae*.

Sub Family
Phocidae are divided into two sub families: the *Monachinae* and the *Phocinae*.
And the *Otariidae* are split in two sub-families: the *Arctocephalinae* (fur seals) and the *otariinae* (sea lions)

Genus
15 genera several being monospecifics.

Species
There is 33 species of pinnipeds: 14 species of Otariidae, 18 species of Phocidae, and one of Odobenidae
The pinnipeds represent 28% of the marine mammals and they number about 50 million, 90% Phocidae, 10% Otariidae (the difference in percentage due to the fact that there are more than 30 million Crabeater seals in the southern hemisphere).

HARBOUR SEAL

Phoca vitulina concolor

ORDER   Carnivores
SUBORDER   Pinnipeds
SUPER FAMILY   Phocoidae
FAMILY   Phocidae
SUBFAMILY   Phocinae
GENUS   Phoca
SPECIES   *Phoca vitulina concolor*
VERNACULAR NAMES   Fr. Veau Marin, Loup Marin

The opening to the choanes is a "V."

DESCRIPTION

Length: 1.4-1.75 m.

Weight: Male 60-125 kg, female: 45-77 kg. It is a rather small seal with a well defined forehead when seen from the side.

Color: Grey to yellowish, with spots of varying size, and darkness, mainly on the back. The belly is generally clearer and more uniform in color.

Skull: Can be distinguished from the Greenland seal (*Pagophilus groenlandicus*), with which only it can be confused, by the V shape of the edge of the internal nares (choanae), and the relatively bigger cheek teeth. The mandible too, is more robust, thicker in adults.

NATURAL HISTORY

The Harbour seal is a coastal species, rarely going far offshore. It is relatively sedentary, and live all year long in the Saint-Lawrence Gulf and Maritime Estuary. Solitary enough when in the water, it is a lot more gregarious at the haul out sites. This reflects more on the fact that good haul out sites are rather rare, than on a really gregarious temperament.

Diving: The study of the dives made by a sample of 11 seals showed that 95% of the dives recorded were done at a depth less than 35 m, and 64% of the dives were less than 4 meters deep, lasting 2.7 to 5 minutes (Lesage, 1999). However the Harbour seal can dive a lot deeper and a lot longer, mainly during the winter (down to 508 m, Shreer & Kovacs, 1997), up to 30 minutes.

Diet: Mainly piscivorous (herring, caplin, sand lance, smelt and plaice), but it can also eat shrimp and squid. It is an opportunistic feeder, and will adjust its diet based on the abundance of potential prey. Pups, during their training period, will eat more small crustaceans and small fish. During the summer in the Marine Estuary of the St. Lawrence, it seems to feed more at sunset and during the night. At this time of the day, some fish species, like caplin, herring and sand lances, swim near the surface.

Reproduction: Sexual maturity: males 5-6, females 3-4 years old. Breeding at the end of the nursing period. (Maybe it is the factor which provokes the weaning of the pup.) Males fight for the control of the females, biting the head, neck and shoulders of their opponent. It is often considered monogamous, but it is probably more the availability of the females and the fact that mating takes place in the water that limit the number of females impregnated by a male, rather than a deliberate choice! As with food, it is probably sexually opportunistic.

Gestation: Apparently 11 month. But the implantation of the blastocyst is delayed 2-3 month so the actual gestation period is 8-9 month.

Whelping: On the ground but also in shallow water if need be. The mother seems able to block the birth of the pup if there is danger, and whelp one or two days later (Lawson & Renouf, 1985). One pup only (there has been cases of twins reported), from mid-May to mid June in the Maritime Estuary (Early May to early June on Sable Island). The pups weigh around 10 kg and measure 75 cm. The first molt takes place in the uterus, but sometimes the pup is born with its lanugo that it will shed 2 or 3 days later. They are able to go in the water as soon as they are born.

Nursing: 22-24 days. The pup triples its weight during the nursing period. (50% fat in the milk). The mother is very protective. It recognizes its pup by its odor, and can adopt another young if its own dies early enough. The females, too small to accumulate enough fat to be able to fast during the nursing period, will have to feed during this period, with or without their pup. They will stay near the haul out sites, which have been chosen for security and for room for the pups, as well as the availability of prey nearby.

Longevity:	Longest recorded is 32 years. In the wild they must rarely live that long.
Molting:	From July to mid-September. The animal spends long periods hauled out, because it has to save its energy. There is much blood in the skin to make the new hair grow as fast as possible. In the water it would lose body heat very fast, in spite of its blubber, if it spent too much time there.
Distribution:	Circumpolar, with distinct populations in East and West Atlantic, and East and West Pacific Total world population more than 500,000. There is a sub-species, *Phoca vitulina mellonae* which live landlocked in Lac aux Loups Marins, in northern Québec
Parasites & enemies:	Several nematods species of parasites. Accidentally caught in fishing gear, Often killed by fishermen who see it as a rival. Not legally hunted in Eastern Canada.
Habitat:	Coastal waters, estuaries
Behavior:	Usually solitary, except at the haul out sites

HARP SEAL

Pagophilus groenlandicus

ETYMOLOGY	"Who love the ice" —from Greenland
ORDER	Carnivores
SUBORDER	Pinnipeds
SUPER FAMILY	Phocoidae
FAMILY	Phocidae
SUBFAMILY	Phocinae
GENUS	Pagophilus
SPECIES	*Pagophilus groenlandicus*
VERNACULAR NAMES	Greenland seal

The opening to the choanes is a "U."

DESCRIPTION

It is a rather small seal, 1.6 meters long and weighting 160 kg. Its name comes from the black oval on the back and the flanks of adults which looks like a harp. Their face is black too. When they are born they have still their white and beautiful thick lanugo, then after the first molt, the back becomes darker, with some lighter spots until they are fully adult, and the "harp" appears.

NATURAL HISTORY

It is a migratory species which can make up to 3,000 km between the whelping grounds in the south and the feeding grounds in the north.
It is a gregarious species seen in huge groups on whelping and molting grounds and smaller groups when they are fishing. After a severe decline in the 1960s, and having had a minimum population of 2 million in 1970, the population has been steadily growing until mid-1990 and appear to be stabilized since 1996 at the upper level of the chronological series recorded since 1960 (MPO science, 2000). In 2007 the population was estimated at more than 6 million.

Diving: Can dive down to 400 m and stay 16 minutes underwater.

Diet: Opportunistic feeder, as are all seals, it is mainly piscivorous, eating fish 10 to 20 cm long. It fishes for caplin and arctic cod in the waters off Newfoundland, Mainly caplin in the St. Lawrence Gulf, but also herring, Atlantic cod, Greenland halibut, redfish and plaice. They can eat crustaceans (shrimp and krill) and mollusks (squid). Total prey consumption in Northwestern Atlantic has gone from 2.5 million tons in 1990 to an estimated 3 million in 1999. It represents only 54% of the total catch by the Harp seal, the rest (46%) is eaten in Canadian arctic and Greenland waters. Even though Atlantic cod represent only a small part of the total diet, it is estimated that Harp seals take 49,000 tons of it in Newfoundland waters and 22,000 tons in the St. Lawrence Gulf (MPO Sciences, 2000)

Vocalizations:	The Harp seal can produce more than 17 different sounds underwater, and barking sounds out of it.
Reproduction:	Sexual maturity 4-5 years old (males and females). In early September, Harp seals leave their northern feeding grounds and swim several thousands km south to whelp on the ice floes. Mating is made in the water rarely on the ice. The female seems to mate with several different males.
Gestation:	Total: 11,8 month, but implantation of the blastocyst is delayed for almost 4 months, so gestation proper is no longer than 7 month.
Whelping:	At the end of February to early April. The pup is 80-85 cm long and weighs around 10 kg. It is born with its white lanugo (which has earned it the name White Coat) that it will lose when weaned. It will remain alone on the ice for 4 weeks before being able to swim and feed itself.
Nursing:	Nursing lasts 12 days, during which the pup will gain 30 to 40 kg. Then the mother abandons the pup to mate and feed intensively in preparation for the molting period. The female may not fast the entire period of nursing.
Longevity:	28-35 years
Molting:	On the ice floes, north of the whelping grounds, 4 to 5 weeks after whelping.
Distribution & population:	There are 3 distinct populations, among which there does not seems to be any interbreeding, even if juveniles have been known to go from one area to another. There is the northwestern Atlantic population (5.2 million) breeding in the St. Lawrence Gulf (Gulf population) and off Newfoundland (Front population). Then there is the Eastern Greenland population (more than 500,000) breeding near Jean Mayen island, and finally the Barents Sea population (more than 2 millions), breeding in the White Sea. The total population of the northern Atlantic exceeds easily 7 million.
Parasites & enemies:	Several species of nematods, orcas and sharks may take their toll, as do accidental catches in fishing gear, and, evidently, man. Humans are the principal predators of this seal, which even if it is not an endangered species, or even

at risk of becoming endangered, is used by all kind of organizations as an example. These groups, using partial or truncated information published by scientists or governments in the countries they are hunted, raise alarms on the effects of hunting on the survival of the species. One has to see, too, the pictures found on the internet sites of those associations: crying baby seals (poor guys, they do not have any lacrymal canal, they have no choice but shed tears as an overflow of the liquid lubricating their eyes) in front of a very bloody cadaver of an adult (always presented as the mother). Blood covered ice, partially decomposed cadavers presented as wasted victims killed just for fun or parts (without any proof that they are not dead from natural causes)... Once more it is meant to make often misinformed spectators react emotionally, and often in a very mercantile way, since they are almost always called upon to generously help "save" the seals from extinction by making a financial gift. I hope that the higher numbers given will cause the reader to reflect upon all that. Yes, we have to protect the seals against an overexploitation, but if we want, we can use this resource without endangering it. And do not forget that, in the end, pollution and brutal climatic changes due to greenhouse effect gases will always have more sad effects than a sensible hunt.

Habitat: High seas and coastal waters during the summer, and ice floes for whelping and molting.

Behavior: Gregarious, not only when molting or whelping but when fishing, too, when they seem to cooperate. I have seen, at the end of the winter near Les Escoumins, large herds of seals that apparently were fishing together. On at least three occasions I have even seen, in August in the waters north of Ile-Verte, more than 50 Harp seals fishing. The presence of seagulls following the seals when they were underwater and coming in to feed just before they would surface did not leave any doubt about their activity at the time.

GREY SEAL

Halichoerus grypus

ETYMOLOGY	Hooked nose pig of the sea
ORDER	Carnivores
SUBORDER	Pinnipeds
SUPER FAMILY	Phocoidae
FAMILY	Phocidae
SUBFAMILY	Phocinae
GENUS	Halichoerus
SPECIES	*Halichoerus grypus*
VERNACULAR NAMES	Grey seal, horse head, Fr. tête de cheval, phoque gris

Jean-Pierre Sylvestre

DESCRIPTION

Length: Male 1.95-2,3 m, female 1.65-1.95 m
Weight: Male 170-310 kg, female 95-105 kg.
Appearance: It is a big seal, with a straight profile and a "Roman nose." It is dark grey, with lighter spots not well defined. Females are generally lighter in color, with darker spots. Sexual dimorphism is well marked. The skull is massive, with large nasal fossae. Nasal bones are almost horizontal, and cheek teeth have 2 roots except the last 2 in the maxillary and the last in the mandibles. Accessory cusps on the cheek teeth are not conspicuous

NATURAL HISTORY

The Grey seal is one of the most easily seen of the seals in the Maritime Estuary of the Saint Lawrence river during summer because it is the more abundant. It can be seen farther off shore than the Harbour seal.

Diving: To 500 meters. Can stay underwater for 32 minutes
Diet: As many seals, the Grey seal is an opportunistic feeder, eating the more abundant prey in its environment. Even if the remains of at least 30 species have been found occasionally in its stomach and feces, most of the time the stomach contents is made of the 2 or 3 species most abundant in the vicinity. It is mainly piscivorous (codfish 15%, herring, caplin, smelt and sand lance 69%, plaice and other flatfishes 10,7%), but can also eat crustaceans and mollusks (shrimp and squid). It has been calculated that they eat between 7.5 and 12.5 kg each day. That means, for the whole population, 600,000 tons of prey, including 21,000 tons of cod, just in the St. Lawrence Gulf (Hammill, 1995). It is thought that 40 years ago, when the population was a lot smaller, they ate only 6,000 tons of cod.
Vocalizations: Grey seals make a lot of noise when they are on haul out sites. When they haul out on the reefs N.E. of Ile Verte, one can hear their eerie moanings 2 or 3 km away, which is pretty impressive, especially in fog.

Reproduction:	Sexual maturity: 65% of the females are sexually mature at the end of their 3rd year (and whelp the next year); 89% at the end of their 5th year. The males attain sexual maturity approximately at the same age, but they have to wait to be fully grown to be able to compete with older seals (8th year). Males fight not for a territory but to have access to a certain number of females. They can copulate with 10 females. The Grey seal is polygynic, In some regions mating can take place in the water instead of on land. It has been shown that in this case, males tend to be monogamous, more because of the difficulty to find females that are more dispersed, than by a deliberate choice! Mating takes place from mid-December until mid-October, depending on the location of the mating sites. The males fast during the period of mating so as not loose their control over "their" females
Gestation:	Total of 11.3 months, including the period while the implantation of the blastocyst is delayed (about 3 months). Gestation proper is 7.9 months.
Whelping:	From mid-December to October, according to the regions where it takes place (end of December –early February on Sable Island, and 2 or 3 weeks later in St. Lawrence Estuary and Gulf). The pup is born with its lanugo, and will molt 3 weeks later at the end of the nursing period. It weighs 16 kg at birth, but will grow rapidly, gaining 1.2 to 2 kg each day.
Nursing:	3 weeks. The milk of the Grey seal is very rich, containing 60% fat. The female fasts during nursing.
Longevity:	Around 35 years
Molting:	The first molting takes place just after weaning, then the molting will take place in June-July for the seals of the St. Lawrence Maritime Estuary and Gulf. Those mating on Sable Island return there for the molt.

Parasites & enemies:	There is some commercial hunting and culling of certain populations. Orcas and sharks can take some juveniles, and they can drown in fishing gear. The Grey seal is host to many parasites, among which the infamous cod worm, *Pseudoterranovae incipiens*. In fact it is just an intermediary host. It contaminates itself by eating cod (which are the final host). The parasites multiply in the gut of the seal, and eggs are eliminated with the feces. The seal contribute to the multiplication and dispersal of the parasite. Eggs produce larvae which will contaminate intermediary hosts which will be eaten by the cod and the cycle runs again. It is host of *Anisakisidae* (Marcogliese, Boily, Hammill, 1996) and lung worms (Hammill, Gosselin, Measure, 1998)
Distribution & population:	The Grey seal is divided into 3 distinct populations. One in Western Atlantic (250,000 around Ile de Sable and 50,000 In St. Lawrence Maritime Estuary and Gulf, constantly growing since 1990), one in Eastern Atlantic (130,000 to 140,000) and finally, one, in bad shape because of the pollution, in Baltic Sea (7500). The total population exceeds 400,000.
Habitat:	Rocky islands, sand beaches, ice floes in shallow water. They prefer waters less than 50 m deep.
Behavior:	They assemble on the haul out sites for mating and molting, and disperse widely after that.

J.-P. Sylvestre

ORES - Ursula Tscherter

HOODED SEAL

Cystophora cristata

ETYMOLOGY	Wearing a bladder on the head
ORDER	Carnivores
SUBORDER	Pinnipeds
SUPER FAMILY	Phocoidae
FAMILY	Phocidae
SUBFAMILY	Phocinae
GENUS	Cystophora
SPECIES	*Cystophora cristata*
VERNACULAR NAMES	Juveniles: blue back. Bladder nose seal

DESCRIPTION

Length:
Weight:
Color:

Skull:

This is a very big seal.
Male: 1.8-3.3 m female: 1.5-2.7 m
Male: 275-400 kg female: 275-375 kg
Variable, usually blue grey on the back with darker fuzzy spots. The young have very fine coats of hair almost blue on the back and yellowish on the belly, and for that reason are called "Blue backs."
More squarish viewed from the top, with a short rosttrum, The cheek teeth are rather small and blunt. When they are 4 years old, the males are characterized by an inflatable sac communicating with the nasal fossae (the "bladder" in the Latin name). When inflated this sac can become as big as a football. When the animal closes one nostril and blows air through its nose it can make part of the nasal septum protrude from the other nostril like a blood red balloon. By inflating and deflating it, ping-like sounds are produced.

NATURAL HISTORY

Diving:

Diet:

Can dive down to 1000 m; some dives can last 50 minutes.
Opportunistic, it eats mainly fish (Greenland halibut, redfish, Atlantic cod, caplin, herring) and some invertebrates (mussels, cephalopods, shrimps). It is estimated that the Hooded seals population of the Gulf of St. Lawrence ate

as much as 5,600-8,400 tons of fish in 1995, including significant quantities of Greenland halibut and redfish. The total allowable catch for Greenland halibut in this area is 4,000 tons and the redfish fishery is currently closed due to low stock levels. Thus, the relatively large amount of these species consumed by Hooded seals would appear of the same magnitude or exceed the commercial harvest (Hammill, 1997). The fish eaten are smaller than the ones fished, and mortality rate of those fish is not known, so the real impact of the Hooded seal on the fish stock is not known, compared to other causes of mortality. More research is needed to be more precise on the subject.

Vocalizations: A variety of sounds. The air sacs participate in the emission of some of those sounds.

Reproduction: Female whelp on ice floes. Mating takes place in the water.

Gestation: Total 11.5 months, but implantation of the blastocyst is delayed for almost 4 months The gestation proper is not more than 7.8 month

Whelping: From March to April, on ice floes. The pup weighs around 25 kg and is 1 m long. When females are whelping, males fight each other for a female. The dominant male will remain on the ice with the mother and pup until it is weaned. Then they mate in the water. After mating a male can try to win other females.

Nursing: The Hooded seal has the shortest nursing period of all the mammals. In consequence, its milk is the richest of all mammals, containing 60 to 70% fat. That allows the pup to double its birth weight during the nursing period, even if it lasts only 4 days! The mother fasts during nursing.

Longevity: 35-40 years

Molting: On the ice in Denmark strait (between Iceland and Greenland) in June-August.

Distribution & population: It is divided in 3 distinct populations. One is from northwestern Atlantic and reproduces in two places, partly in the St. Lawrence Gulf (Gulf population) and partly on the Front off

407

Newfoundland (Front population). The second distinct population whelps and mates on the ice of Davis Strait, and the third on the ice east of Greenland, near Ile Jean Mayen. Nothing is known about the possibility of interbreeding between those 3 populations. There is at least one case known of hybridization between a Harp seal and a Hooded seal.

Population: More than 650,000. It is estimated.

Parasites & enemies: Sharks and orcas can kill some juveniles

Habitat: The high sea and on ice floes

Behavior: The sacs, when inflated, are used to demonstrate aggressiveness when threatened or during rutting. By inflating and deflating them it can produce several different sounds. It makes long migrations between the whelping and molting sites, and the places where it feeds.

WALRUS

Odobenus rosmarus

ORDER	Carnivores
SUBORDER	Pinnipeds
SUPER FAMILY	Phocoidae
FAMILY	Odobenidae
GENUS	Odobenus
SPECIES	*Odobenus rosmarus*
VERNACULAR NAME	aikivik (Inuit)

DESCRIPTION

Length & weight: It is, with the Elephant seal, the biggest of the pinnipeds. Males can weight as much as 1,750 kg for a length of 3.6 m, and the females, a little smaller 750 kg for 2.5 m, for the Pacific subspecies *O.r. divergens*. The other subspecies, *O.r. rosmarus*, from the north Atlantic, and *O.r. laptevi*, living north of Siberia in the Laptev Sea (this one is not always considered as a subspecies but as a separate population of *O.r. divergens*) are a little smaller.

Color: It is cinnamon colored, but this color lightens with age and the skin temperature. When a Walrus has spent a long time in the water, its skin (which can be 4 cm thick in the neck and shoulder region) becomes almost white. It is because the blood has left it, circulating mainly under the blubber to limit loss of heat. However, when the Walrus has been out of the water for some time, and it is sunny, its skin becomes brick red, because lots of blood circulates in it to lower the body core temperature. Its skin looks naked, but in fact is covered with short, sparse, and coarse hair. Males and females have long canines in the upper maxillary, which can be more than a meter long in males.

Skull: The skull is massive and is easily recognized by the huge teeth sockets holding the roots of the canines. The mandible syphysis is welded. Walrus can bend their feet under their body when on land.

NATURAL HISTORY

The Walrus prefer shallow water, rich in the shellfish that are the main part of their diet. They dig out the mollusks by spitting powerful jets of water into the sediments, then sucking them into their mouth. They crush the smaller ones between their flat cheek teeth. The bigger ones are extracted from their shell, by powerful suction. The very strong muscles of the pharyngeal region make a powerful piston of the tongue, creating negative pressure in the mouth of more than one bar (one bar is equivalent to one atmospheric pressure). It is quite enough to extract the animal from its

J.-P. Sylvestre

shell. Walrus follow the ice movements. When winter comes and the ice is moving south, so do they. In the spring when the edge of the ice moves back to the north, they follow. They can migrate for distances as great as 3,000 km Males seem to prefer rocky shores and islands; females prefer the edge of the ice. Canines are used as for power displays by the males, and to defend themselves and to climb on the ice by both sexes.

Diving: The Walrus is surely able to make deep dives (100 m Shreer & Kovacs, 1997) but most of the time it will dive less 35 m, spending 15 minutes or so underwater.

Diet: Mainly shellfish, but can also kill seals and suck the blubber from them.

Vocalizations: Very vocal all the time but mainly during the rutting season. In the neck region, it has big sacs that can be inflated to create buoyancy in the water, and which can act as resonating chambers when it vocalizes.

Reproduction: The male Walrus fight to keep exclusive access to a group of females. Males are not necessarily limited to only one group of females during the mating season, which takes place in January and February. They mate in the water.

Gestation:	Total 15 months, but the implantation of the blastocyst is delayed 4.5 months, so that the gestation proper does not exceed 10.5 months.
Whelping:	From mid-April to mid-June. One pup only, 1 m long, weighing 63 kg. It can swim at birth.
Nursing:	More than 2 years, but the young can feed by themselves before weaning. As one can infer from the length of the nursing period, the milk is a lot less rich than in phocids (only 30% fat).
Longevity:	More than 40 years
Distribution & population:	Circumpolar. Divided into 3 units: one in north Atlantic, one in eastern Pacific, and one in the Laptev Sea, which is considered by some scientists as a subunit of the Pacific unit. The population total is in excess of 300,000, which does not place the Walrus among the endangered species.
Parasites & enemies:	Polar Bears can kill some juveniles, but humans are the main predators. Aborigines hunt Walrus in "survival" hunts. Walrus has been eaten used as food and dog food by aborigines. Its skin makes good thongs and its teeth have always had value, as raw material for tools, harpoon heads, sled runners, carvings, etc. by aborigenes, and as an ivory source for Europeans and Asians. Canines have kept a good value, and are the cause of some poaching. Old hunting methods employed a harpoon with a float, which allowed for the recovery of the killed animal. New hunting methods use high power rifles, often from far away, causing the loss of killed animals, which are never accounted for. So it is very difficult to establish quotas or to evaluate the real pressure exerted on the species by aboriginal hunting, the only type allowed today. A lot more menacing to the survival of the species, however, are the projects for the exploitation of gas and oil resources in the general area occupied by the Walrus. People want to exploit also the shellfish that is peculiarly abundant there. Taking in account the effects on the seabed of such fisheries,

(the bottom is scoured clean by heavy trawls or powerful jets of water, and what is not taken is crushed) we can really begin to worry about the future of the Walrus. Anti-hunters campaigns can be easily organized, just by using emotional impact caused by the misuse of firearms or the apparent cruelty of the hunt, but when money speaks...

Habitat: Shallow seas, rich in shellfish, along the edge of the ice

J.-P. Sylvestre

NORTHERN ELEPHANT SEAL

Mirounga angustirostris

ORDER	Carnivores
SUBORDER	Pinnipeds
SUPER FAMILY	Phocoidae
FAMILY	Phocidae
SUBFAMILY	Monachinae
GENUS	Mirounga
SPECIES	*Mirounga angustirostris*

The canines are very big, because they are used in dominance-establishing fights.

DESCRIPTION

It is the biggest of all pinnipeds, males weighing 2300 to 2500 kg. The boreal species (M.leonina) can even grow bigger, with males weighting up to 3500 kg. It is also the species showing the maximum sexual dimorphism, the female rarely exceeding 600 kg.

Sexual dimorphism is evident on this picture (Piedras Blancas, California)

The male has a large proboscis used in dominance (visual and auditory) displays with other males and responsible for its vernacular name. Males have also a kind of shield on the neck region and the chest made from scar tissue thickening after each breeding season. Females do not have a proboscis, and neither do juvenile males. It begins to grow near sexual maturity, (3-5 years) and will be fully developed when the males are between 7 and 9 years old.

The proboscis; the shield made of scar tissue can be seen, too.

NATURAL HISTORY

Boreal Elephant seals are among the animals making the longest migrations between the rookeries and the feeding grounds. Males can make 10,000 km round trips twice a year. Males and females feed actively during the month spent at sea to replenish the blubber that will be used during the mating season and the molt, when they are fasting. They spend more than 80% of their annual cycle feeding at sea.

Diving: Elephant seals break some marks here too as well for the depth than the time spent underwater.
- Males: maximum depth recorded 1567 m . Longest apnea: 77 minutes
- Females: maximum depth recorded: 1273 m. longest apnea : 62 minutes

However, their usual dives do not seem to exceed their aerobic dive limit (ADL) since they seem to dive continuously staying only 2-3 minutes at the surface. Apnea then rarely exceeds 20-30 minutes and depth are around 400 m for the females and 750-800 m for the males. This seems to show that they do not forage in the same place.

Diet: Mainly pelagic fish and squid. They can also take some prey near the bottom when the depth allows it.

Vocalizations: Snoring, barking, etc. The end of the proboscis hangs in the open mouth which acts as resonance chamber. The females make sounds halfway between a bark and a grunt. The pups have a clearer bark so they can be recognized by their mother.

Reproduction: The Elephant seals, Northern as well than Southern are polygynic. The males arrive on the breeding grounds in December, and defend a portion of beach where they establish a kind of hierarchy.

Male in challenge posture. The open mouth acts as a resonance chamber, amplifying the grunts of the animal.

The females arrive a little later and form groups or a harem around a male, who will chase away any rival by using threatening postures and vocalizing very loudly (it is a way to keep their energy for the fights with other males that have not been intimidated by the oral and physical displays. They whelp almost immediately and begin suckling their pups, giving them a milk with a rather low fat content (10%) at the beginning, but becoming very high (54%) as soon as the first quarter of the lactation period. This allows the female to save body fluids, since it will fast until the pup is weaned.

Piedras Blancas haul out site, near San Simeon, California.

They come into oestrus around weaning time and are mated by the dominant males. Then they take to the sea, letting the pups fend for themselves. The fighting between males becomes ferocious and virtually only alpha males will mate (On Año Nuevo Islands, it is estimated that only 5 males are responsible for 45 to 95% of the matings.) Males that do not succeed in establishing their dominance will not mate. It is easy to understand the importance of the sexual dimorphism. The bigger the male the more chance it has to become dominant.

Principal haul out sites of the Southern Sea Elephant on the west coast of America.

Gestation: 11 months, with a delayed implantation of the blastocyst of 4 months.

Whelping: December to January. The pup weight is 40 kg when born. It will weigh more than 150 kg when weaned.

Nursing: 26-28 days. The mother stays with the pup, fasting the whole period. The milk contains up to 54% fat, the pup will take 4 kg a day. Some pups when weaned, will try to suck other females. If they succeed they will multiply their

The happy pup finds the udders full.

weight at birth by 7. It will give them a better chance to survive, often at the expense of the young of the mother they steal the milk from. The females can lose 4 kg of body mass for each kg gained by the pup. Once the pup is weaned, it will spend 4-6 weeks alone without eating, living on the blubber it has accumulated, learning to swim and to feed itself. It is a very hard period and many die during this time. Then they take to the sea for 6 months and many more will die (up to 30%) before they come back to the haul out sites in September-October.

Enough for everyone.

419

Longevity:	Males 17 years, females 22 years.
Molting:	Elephant seals have an unusual molt. Not only is the hair shed, but part of the epidermis, too. They have to come back to the haul out sites because it would be too costly to maintain a constant body temperature, in spite of the need to bring a the rich supply of blood to the skin, were they to remain in the water. They will fast again at this time. A male can lose more than 700 kg during fasting periods.
Distribution & population:	All along the western coast of America, from Baja California to Alaska
Parasites & enemies:	**Natural:** several kinds of nematods, sharks, and Orcas **Humans:** Humans almost exterminated the Elephant seal. A big male could give up to 100 liters of oil, so it was looked for by hunters. During the 1880s, 7 of what was thought to be the last 8 Elephant seals were killed on Guadalupe Islands. There must have been some more somewhere, because more and more were seen and now the population exceeds 150,000 individuals and seems to increase by 6% each year. They are completely protected in USA. Some are caught as bycatch in some fishing gear, but climatic changes (the El Niño effect), by increasing the frequency and the violence of winter storms, causes the death of many pups, and interferes with mating. Their consanguinity does not seem to affect their survival rate.
Habitat:	Protected sandy and rocky beaches and islands during the mating and molting season, and at sea on the edge of the continental shelf the rest of the time (250 days) when they feed to replenish their fat stores.

Establishing the hierarchy. The male on the left, younger if we consider the size of the proboscis, was not able to keep its ground very long.

Advanced courting.

Copulation takes several minutes.

Each birthing gives happiness to the gulls which feed on the afterbirth.

The females put sand on their back, maybe to protect themselves from the sun.

Bachelors beach. Juvenile males too big to be tolerated on the main haul out site have found a quiet beach to rest.

Young male beginning its training. The cause of the fight is easy to guess.

Young male wearing the marks left by its training.

GLOSSARY

ACOUSTIC IMPEDANCE
Resistance to sound transmission.

ADIABATIC RELEASE
A sudden decrease in the pressure of a gas caused by an increase in its volume without loss or gain of heat.

ANAEROBIC METABOLISM
Metabolic processes that occur without oxygen.

ANASTOMOSIS
Interconnection between two blood vessels.

ANTHROPOCENTRISM
The explanation of everything in terms of human experience or values; the view that the human being is the center of the universe.

ANTHROPOMORPHISM
The tendency of attributing human characteristics to things or animals.

APNEA
A temporary suspension of breathing.

APOPHYSIS
A projection from a bone, usually for muscle attachment.

ARCHAEOCETES
Organisms that flourished in the Eocene-Miocene epochs highly adapted for the aquatic life, but exhibiting characteristics of their terrestrial ancestors. The archaeocetes show similarities with the present-day cetaceans (such as the internal ear) as well as archaic characteristics. They are probably part of the link leading to the recent species.

ARRECTOR PILI MUSCLE
A muscle located at the base of a hair. It is attached to one side of the follicle so that its contraction pulls the hair upright. Its contraction also causes human goose bumps and the fluffing of bird feathers or animal fur.

ARTERY
A blood vessel that carries oxygenated blood away from the heart.

ARTICULATION
The point of attachment (joints) of two or more bones. An articulation may be movable (elbow) or immovable (bones of the skull).

ARTIODACTYLS
Even-toed ungulate mammals which include animals such as the hippopotamus, the pig, and the ruminants.

AUTOTROPH
An organism that makes its own food from the inorganic materials in the environment. Some organisms use light energy for the synthesis (i.e., photoautotrophs), while others use energy derived from chemical processes (i.e., chemoautotrophs).

BACULUM or OS PENIS
The penis bone of certain mammals (seals, weasels, dogs, bears, etc.).

BARORECEPTOR
A receptor that responds to changes in pressure.

BENTHIC
Aquatic organisms that live on the bottom of a sea or lake, and depend on it for their survival.

BIOCENOSIS
A self-sufficient community of organisms occupying and interacting within a specific biotope.

BIOTOPE
An ecosystem with its own particular community.

BRACHIOSAURUS ("arm lizard")
One of the largest and tallest dinosaurs. This gigantic herbivore lived in the Late Jurassic Period.

BRADYPODIDAE
Arboreal mammals belonging to the order Xenarthra. Also named "Tree sloths," they live in South America.

CAPILLARIES
Microscopic blood vessels that conduct blood from the arterioles to all living cells and waste material to venules for ultimate excretion.

CARPI
The wrist bones.

CEPHALOPODS ("head-foot")
Marine mollusks that have a circlet of prehensile tentacles and a mouth equipped with a chitinous beak. Architeutis spp. and the giant squids are cephalopods.

CERVIX
The narrow or necklike muscle of the uterus leading to the vagina of mammals.

CETACEANS
Streamlined, almost hairless, and entirely aquatic mammals whose forelimbs have been modified to form fins and whose hind limbs are absent or vestigial. The tail (caudal fin) is horizontal and lacks skeletal elements. The nasal openings (blowholes) are located on the dorsal surface of the skull.

CETOTHERIDAE
Primitive Mysticetes that lived between the Middle Oligocene and Early Pliocene. They probably had rudimentary baleen plates instead of teeth.

COCHLEA
In the internal ear, part of the labyrinth that contains the auditory organs.

COLLAGEN
A complex protein that makes up the intercellular substance of a tissue.

CONDUCTION
The action of transmitting heat through physical contact.

CONES
A type of light-sensitive receptor cell in the retinas of vertebrates. They convey information about color and work best in bright light.

COPEPODS
Small crustaceans, usually 0.5 to 2 mm long, that occur in marine and freshwater habitats. They lack a carapace and are important members of the plankton.

CORIOLIS FORCE
The force resulting from the rotation of the earth. It affects the atmosphere and ocean currents.

CORPUS ALBICANS
The white scar resulting from the degeneration of the corpus luteum.

CORPUS LUTEUM
The transformation of the follicle following ovulation. It functions as a temporary endocrine gland secreting progesterone.

CREODONTS
The oldest carnivorous mammals. They appeared in the Early Paleocene epoch (55 to 65 million years ago). Dentition is complete and typical of carnivores.

DEMERSAL
Animals, such as fish, that live close to the sea bottom.

DISTAL
Farther from the midline or point of attachment to the trunk.

ECOLOGICAL NICHES
The functional position occupied by an animal in its biocenosis. It is the equivalent of the workplace in the human community.

ECOSYSTEM
A group comprising many species (biocenosis) occupying and interacting within a determined space (biotope).

ELECTRON TRANSPORT CHAIN
A sequence of biochemical reactions important in aerobic respiration. These reactions occur within the mitochondria (animals) and chloroplasts (plants).

EOCENE
   The second geological epoch of the Tertiary period (60 million years ago). It precedes the Oligocene, the epoch in which whales appeared.

EPIDIDYMIS
   Part of the male genitalia involved in the reception and maturation of the spermatozoa.

ETHMOID
   A bone which forms the upper part of the nasal cavity. The cribriform plate is penetrated by the olfactory nerves.

EXTERNAL EARS or *PINNAE*
   Located on either side of the head, often mobile, they capture sounds.

FASCINE FISHING
   A type of stationary fishing gear that resembles a large circular trap used on the shores of the St. Lawrence River.

FOLLICLE
   A small fluid-filled spherical cavity in the ovary that contains the oocyte.

FOSSILS
   The remains of a once-living organism preserved in the sedimentary deposits of the earth's crust.

FRONTAL
   The bone that forms the forehead and the roof of the orbit.

GLENOID CAVITY
   The socket into which the head of a bone fits.

GRAVIPORTAL
   Applied to limbs that are usually short and massive serving as vertical columns.

HERTZ
   The SI unit of frequency (symbol: Hz), equal to one cycle per second.

HETEROTROPH
   An organism that feeds on organic compounds manufactured by other organisms. Included among the heterotrophs are all the animals and the nonautotrophic plants.

HOMEOTHERM
An organism whose body temperature is maintained within narrow limits. Opposite to poikilotherm.

HUMERUS
The upper bone of the forelimb of tetrapods. It is located between the shoulder and the elbow.

HYDRODYNAMIC
Pertaining to the way a liquid flows along a body or to the resistance that it generates against the movement of that body.

HYDROSTATIC PRESSURE
The pressure exerted by water or other fluids on a body.

HYOID APPARATUS
Gk. hyoeides: shaped like the letter upsilon (u). A bony structure located at the base of the tongue of vertebrates.

ICHTHYOSAURS "fish lizards"
Dolphin-like reptiles with fins and a vertical tail that lived during the Mesozoic, between 190 and 70 million years ago.

INDRICOTHERIUM
A hornless form of rhinoceros that lived in Asia during the Oligocene. The largest land mammal ever. It measured more than 5 m (18 ft) high and weighed perhaps 20 tons.

INTERVERTEBRAL DISKS
Fibrocartilaginous disks separating two adjacent vertebral bodies (or centra).

KERATIN
A scleroprotein forming the structural bases of hair, hoofs, feathers, claws, nails, and baleen plates.

KRILL
Norwegian word meaning whale food. Many shrimp-like crustaceans are referred to as krill (Euphausia sp., Thysanoessa sp., etc.).

LABYRINTH
The structural elements of the internal ear that include the semicircular canals and the cochlea.

LACTIC ACID
Metabolic product that appears in active muscle tissue when oxygen is limited.

LENS
A biconvex structure behind the eye pupil that focuses images onto the retina.

LEVIATHAN
Marine monster of Phoenician mythology mentioned in the Bible.

LOOP of HENLE
The part of the nephron (the filtration unit of the vertebrate kidney) responsible for the extraction of water and some salts during the production of urine.

MAXILLARIES
The paired posterior bones of the upper jaw in vertebrates. They carry all the upper teeth except the incisors.

MECHANORECEPTOR
A receptor that responds to mechanical stimuli. They may react to touch, pressure (baroreceptor), or to sound.

MELON
A waxy lens-shaped structure in the forehead of toothed whales.

MENISCUS
A crescent-shaped fibrocartilaginous disk found between joints.

MILLISECOND
One-thousandth of a second.

MIOCENE
An epoch of the Tertiary era dating back 25 million years ago and characterized by the appearance of many mammalian species.

MYOCARDIUM
The thickest part of the heart wall. It is composed of cardiac muscle.

NEKTON
Aquatic organisms that can swim against currents, as opposed to plankton.

NEMATODES
A group of cylindrical and usually parasitic worms.

NEREIS "clam worms"
Marine polychaetes characterized by having chitinous jaws, which they use to cut their food, and rudimentary limbs called parapodia, which also play a role in gas exchange. They are often used as fishing bait.

NEURAL or *SPINAL PROCESSES* (neural spines)
   Bony extensions that project laterally from the vertebral, or neural arch surrounding the spinal cord.

OCCIPITALS
   The bones that form most of the floor of the posterior part of the head and serve as both articulation and anchorage within the neck.

OLFACTORY EPITHELIUM
   Epithelium of the nasal mucosa having olfactory receptors.

OLIGOCENE
   An epoch of the Tertiary Period starting about 25 million years ago and ending approximately 5 million years ago.

ORGAN of CORTI
   The cochlear organ that converts sounds to nerve impulses. It is responsible for our hearing.

OTOLITHS (ear stones)
   Calcareous secretions in the ears of many animals. They play a role in spatial positioning and movement. Aside from their shape being unique to each species, they are often very large in fish and can be used in the determining their age as well as in establishing the diet of piscivores (we find them in their stomach or feces).

PALEONTOLOGY
   The science of the forms of life that once existed on earth. It deals with the study of fossils.

PARIETALS
   The bones that make up most of the walls and roof of the braincase.

PATRIOCETIDS
   Extinct marine animals that lived during the Late Eocene and Late Oligocene epochs. Among these animals are Patriocetus and other similar types that possibly represent intermediate forms between the archaeocetes and Mysticetes.

PELAGIC
   A term which refers to organisms inhabiting open water, as opposed to benthic.

PHALANGES
   In vertebrates, the bones that make up the digits of the hand or foot.

PHEROMONES
   Odiferous substances emitted by many animals into the environment that convey chemical messages.

PHOTOSYNTHESIS
   A chemical process in plants equipped with photosynthetic pigments such as chlorophyl responsible for manufacturing organic compounds from carbon dioxide and water in the presence of sunlight.

PHYSTERIDAE
   Cetacean family that includes the Sperm whale, the Pygmy Sperm whale, and the Dwarf Sperm whale.

PHYTOPLANKTON
   The plant component of plankton.

PLACENTAL (Eutheria)
   Mammals whose fetus develops entirely within the uterus nourished by the placenta.

PLANKTON
   Aquatic organisms that are unable to swim effectively against water currents, as opposed to nekton.

PLIOCENE
   An epoch of the Tertiary Period from approximately 5 to 2 million years ago.

POIKILOTHERM
   An organism whose body temperature fluctuates with the one of its surroundings.

POLYPHALANGIA
   The presence of additional phalanges (more than 3) in the digits of the hand or foot

PREHENSILE
   Having the ability to seize, grasp, or hold on.

PREMAXILLARIES
   The bones of the head that bear the incisors.

RETE MIRABILE ("Marvelous net")
   A network of anastomosed veins and arteries that can serve as a blood reservoir.

RODS
: A type of light-sensitive receptor cell in the retinas of vertebrates. They convey information about monochromatic vision and work best in dim light.

SCAPULA (shoulder blade)
: The skeletal element of the pectoral girdle that connects the forelimb to the trunk.

SEBACEOUS GLANDS
: A cutaneous gland that develops in the hair follicle and secretes oily and waxy materials.

SEROLOGICAL
: Pertaining to the testing of blood serum.

SESSILE
: A term that describes animals that live permanently attached to a surface.

SIRENIANS
: A group of herbivorous marine mammals related to elephants. They include manatees and dugongs.

SQUALODONTIDS
: A group of extinct marine mammals similar to dolphins. Their teeth were differentiated into incisors used in seizing prey and acerated molars resembling those of sharks. They lived during the Middle Oligocene and Middle Pliocene epochs. They possibly represent the ancestors to the present-day Odontocetes. The nostrils were located at the top of the head.

STATORECEPTORS
: Receptors that convey information to the nervous system. They are associated with the control of movement and posture.

STERNAL RIBS
: Bony structures connecting the ribs to the sternum.

SUPER-PREDATORS
Animals that feed on various levels of the food chain. They feed on primary producers, secondary, and tertiary consumers.

SYNOVIA
A lubricating liquid produced by certain membranes of joints.

TAPETUM LUCIDUM
A reflective layer located within or behind the retina of some vertebrates. It is associated with nocturnal or crepuscular animals. It increases light sensitivity.

TETRAPODS
Any four-legged animal.

TRANSVERSE PROCESSES
Lateral bony projections from the vertebral body with which the ribs and other vertebrae articulate.

TROPHIC LEVEL
The position that an organism occupies in the food chain (e.g., primary producers, primary consumers, secondary consumers, etc.).

VAS DEFERENS
One of the paired ducts that transport sperm from the testis to the copulatory organs.

VEINS
Blood vessels that carry blood to the heart.

VERTEBRAL BODIES
The most important part of a vertebra. It is generally spool-shaped.

VAGINAL PLUG
A structure found in the vagina of certain Odontocetes.

VIBRISSA
A tactile hair of certain mammals.

ZOOPLANKTON
The animal component of plankton.

Bibliography

AGUILAR, A., L. JOVER and E. GRAU, 1981, "Some anomalous disposition of the Jacobson organ," *Scientific Reports of the Whale Research Institute*, vol. 33, pp. 125-126.

AMOS, B., C. SCHLÖTTERER and D. TAUTZ, 1993, "Social structure of pilot whales revealed by anatical DNA profiling," *Science*, vol. 260 (April), p. 340.

ANDREASE, C. et al., 1988, *Our Way of Whaling-Arfanniariaaserput*, Greenland Home Rule Authority, May, 30 p.

ASHLEY, C.W., 1991, *The Yankee Whale*, Dover Publications Inc., (first published in 1926, reprinted in 1942), 143 p.

AU, W.L., 1993. *Sonar of Dolphins*, Springer-Verlag.

BAKER, R.R., J.G. MATHER and J.H. KENNAUGH, 1993, "Magnetic bones in human sinuses," Letters to Nature, *Nature*, vol. 301 (January 6), p. 78-80.

BATEMAN, G. et al., *Les mammifères marins*, Paris, France Loisirs, 143 p.

BEAMISH, P. and E. MITCHELL, 1971, "Ultrasonic sounds recorded in the presence of a blue whale *Balaenoptera musculus*," *Deep Sea Res* 18, pp. 803-809.

BÉLAND, P., 1996, *Le béluga, adieu aux baleines*. Montreal, Libre Expression.

BÉLAND, P., D. MARTINEAU, P. ROBICHAUD, R. PLANTE and R. GREENDALE, 1987, *Échouages de mammifères marins sur les côtes du Québec dans l'estuaire et le golfe du Saint-Laurent de 1982 à 1985*, Rapport technique canadien des sciences halieutiques et aquatiques, no. 1506.

BENTON, M.J., 1990, *Vertebrate Paleontology Biology and Evolution*, London, Harper Collins Academic.

BEUCHAT, C.A., 1996 "Mammalian kidney: A correlation with habitat," *Am J Physiol* 271 (Regulatory Integrative Comp. Physiol., 40), pp. 157-179.

BONNER, N., 1989, *Whales of the World*, New York, *Facts on File*, 191 p.

BOVET, J., M. DOLIVO, C. GEORGE and A. GOGNIAT, 1988, "Homing behaviour of wood mice (*Apodemus*) in a geomagnetic anomaly," *Z. SäugertierKunde* 53, pp. 333-340.

BRETON, M., *Guide d'observation des baleines au Canada*, Department of Fisheries and Oceans Canada, 53 p.

CARWARDINE, M., 1995, *Baleines, dauphins et marsouins*, collection L'œil Nature, Paris, Bordas, 255 p.

CAVE, A.J.E. and F.J. AUMONIER, 1962, "Morphology of the *cetacean reniculus*," *Nature*, no. 4817 (February 24), pp. 799-800.

CLEAVE, A., 1994, *Whales and Dolphin: A Portrait of the Animal World*, New York, Todtri, 80 p.

COLLECTIVE WORK, 1972, *La baleine*, Stock, 287 p.

_____, 1985, "The sirenians and baleen whales,, in S. RIDGWAY and R. HARRISON (dir.), *Handbook of Marine Mammals*, vol. 3, Academic Press, 362 p.

_____, 1986, "About cetaceans: An update in waters," *Journal of Vancouver Aquarium*, vol. 9, 36 p.

_____, 1989, "River dolphins and the larger toothed whales," in S. RIDGWAY and R. HARRISON (ed.), *Handbook of Marine Mammals*, vol. 4, London, Academic Press, 442 p.

COFFEY, D.J., 1977, *The Encyclopedia of Sea Mammals*, London, Hart-Davis, MacGibbon, 223 p.

CORRIGAN, P., 1991, *Where the Whales Are Your Guide to Whale Watching Trips in North America*, Chester (CT), The Globe Pequot Press, 326 p.

COUSTEAU, J.Y. and Y. PACCALET, 1986, *La planète des baleines*, Paris, Robert Laffont, 280 p.

COX, V., 1989, *Baleines et dauphins*, Courbevois, Soline, 128 p.

CRANFORD, T., M. AMUNDIN and K.S. NORRIS, 1996, "Functional morphology and homology in the Odontocete nasal complex: Implication for sound generation," *Journal of Morphology* 228, pp. 223-285.

DIETZ, T., 1983, *Tales of the Sea*, Portland (ME), Guy Gannet Publishing Co., 100 p.

DOSIER, T.A., 1977, *Whales and Other Sea Mammals*, Time Life, 128 p.

DOW, G.F., 1985, *Whaleships and Whaling: A Pictorial History*, New York, Dover Publications, (Reprinted from the edition published by The Marine Research Society, Salem, MA, 1925), 241 p.

DOW, L., 1990, *Whales*, New York, *Facts on File*, New York, 67 p.

DU PASQUIER, T., 1990, *Les baleiniers français de Louis XVI à Napoléon*, Paris, Henry Veyrier et Kronos, 227 p.

ELLIS, R., 1989, *Dolphins and Porpoises*, New York, Knopf, 270 p.

_____, 1991, *Men and Whales*, New York, Knopf 1991, 542 p.

_____, 1994, *The Book of Whales*, New York, Knopf, 202 p.

FISH, F.E. and C.A. HUI, 1991, "Dolphin swimming: A review," *Mammal Rev*, vol. 21, no. 4, pp. 181-195.

FONTAINE, P.-H. 1995, "Échouage d'une baleine à bec sur les battures de Montmagny le 6 novembre 1994," *Le Naturaliste canadien*, vol. 119, no. 2 (Summer), pp. 48-53.

_____, 1996, "De la présence de dents vestigiales non fonctionnelles chez certains cétacés," *Le Naturaliste canadien*, vol. 120, no. 2 (Summer), pp. 34-38.

_____, 1997. "Comparaison de la cavité buccale et de la région pharyngienne des odontocètes et des mysticètes," *Le Naturaliste canadien*, vol. 121, no. 2 (Summer), pp. 20-24.

FONTAINE, P.-M. and C. BARRETTE, 1997, "Megatestes: Anatomical evidence for sperm competition in the harbour porpoise (*Phocoena phocoena*)," *Mammalia*, vol. 61, pp. 65-71.

GALANTSEV, V.P.,1991, "Adaptational changes in the venous system of diving mammals," *Can J of Zoology*, vol. 69, pp. 414-419.

GARDNER, R., 1946, *The Whalewatcher's Guide*, Messner, 170 p.

GASKIN, D.E., 1972, *Whales, Dolphins and Seals*, Hennerman Auckland, 459 p.

GERRIT, S.M. Jr., "The telescoping of the cetacean skull," *Smithsonian Miscellaneous Collections*, vol. 76, no. 5.

GIACOMETTI, L., "The skin of whale (*Balaenoptera physalus*)," *Anat Rec* 159, pp. 69-76.

GOOLD, J.C. and S.E. JONES, 1995, "Time and frequency domain characteristics of sperm whale clicks," *J Acoust Soc Am* 98, no. 3 (September), pp. 1279-1291.

GRASSÉ, P.P. and C.H. DEVILLIERS, 1965, *Zoologie II. Vertébrés*, Paris, Masson et Cie.

HEINYRICH, D. and M. HYERGT, 1993, *Atlas de l'écologie*, Paris, La Pochotèque, 284 p.

HERMAN, L.M. et al., *Cetacean Behavior: Mechanism and Functions*, Interscience Publications, 257 p.

HILYDEBRAND, M., 1982, *Analysis of Vertebrate Structure*, New York and Toronto, John Wiley & Sons, 654 p.

HOYT, E., 1985, *The Whale Watchers's Handbook*, Toronto, Penguin Madison Press Book, 208 p.

_____, 1990, *Orca, the Whale Called Killer*, Camden East, Camden House, 291 p.

KALYMAN, B. and K. FAYRIS, 1988, *Arctic Whales and Whaling*, The Arctic World Series, Crabtree Publishing Co., 57 p.

KATYONA, S., D. RICHARDSON and R. HAZARD, 1977, *Whales and Seals of the Gulf of Maine*, second edition, Bar Harbor (ME), College of the Atlantic, 100 p.

KEMPF, H., *La baleine qui cache la forêt. Enquête sur les pièges de l'écologie*, Paris, La Découverte, 1994, 220 p.

KING WINN, L. and H.E. WINN, 1985, *Wings in the Sea: The Humpback Whale*, Hanover and London, University Press of New England, 151 p.

KIRSCHWINK, J.L., A.E. DIZON and J.A. WESTPHAL, 1986, "Evidence from strandings for geomagnetic sensitivity in cetaceans," *J Exp Biol* 120, pp. 1-24.

KLINOWSKA, M., 1986, "The cetaceans magnetic sense: Evidence from stranding," in M.M. BRYDEN et R. HARISON (ed.), *Research on Dolphins*, pp. 401-432

LAMBERTSEN, R., N. ULRICHET and J. STRALEY, 1995, "Fronto-mandibular stay of *balaenopteridae*: A mechanism for momentum recapture during feeding," *Journal of Mammalogy*, vol. 76, no. 3, pp. 877-899.

LAURIN, J., 1982, "La chasse au béluga (*Delphinapterus leucas*) du Saint-Laurent et statut actuel de la population 1982," *Les Carnets de Zoologie*, vol. 42, no. 2, pp. 23-27.

LEATHERWOOD, S. and R.R. REEVES, 1983, *Whales and Dolphins*, Sierra Club Handbook, 302 p.

LEATHERWOOD, S., D.K. CALDWELL and H.E. WINN, 1976, "Whales, dolphins and porpoises of the Western North Atlantic: A guide to their identification," NOAA technical report NMFS CIRC-396, Seattle (WA).

LIEN, J., 1985, *Wet and Fat Whales and Seals of Newfoundland and Labrador*, St. John's (NF), Breakwaters Book, 136 p.

LIEN, J. and S. KATONA, 1990, *A Guide to the Photographic Identification of Individual Whales Based on Their Natural and Acquired Markings*, American Cetacean Society, St. John's (NF), Breakwater.

LILLY, J.C., 1962, *L'homme et le dauphin*, Paris, Stock, 199 p.

LILLY, J.C. and A. MILLER, 1961, "Vocal exchanges between dolphins", *Science* 134, pp. 1873-1876.

LOUGHLIN, T.R., L. CONSIGLIERI, R.L. DELONG and A.T. ACTOR, 1983, "Incidental catch of marine mammals by foreign fishing vessels 1978-81," Seattle, *Marine Fisheries Review* 45, pp. 7-9.

LYNAS, E.M. and J.-P. SYLVESTRE, 1988, "Feeding techniques and foraging strategies of Minke whales (*Balaenoptera acutorostrata*) in the St. Lawrence River estuary," *Aquatic Mammals*, vol. 14, no. 1, pp. 21-32.

MADDEN, R.C., 1987, "An attempt to demonstrate magnetic compas orientation in two species of mammals," *Animal Learning and Behaviour*, vol. 15, no. 2, pp. 130-134.

MARTIN, K., 1988, *Giants of the Sea*, New York, Gallery Books, 37 p.

MARTIN, R., 1978, *Les mammifères marins*, Elsevier, 206 p.

MATTHEWS, H.L. 1978, *The Natural History of the Whale*, Columbia University Press, 219 p.

McGHEE R., 1988, "Beluga hunters: An archaeological reconstruction of the history and culture of the Mackenzie Delta Kittegaryumiut," Canadian Museum of Civilisation, *Newfoundland Social and Economic Studies*, no. 13, 124 p.

MEASURES, L.N., 1992, "Bolbosoma turbinella (*Acanthocephala*) in a Blue whale (*Balaenoptera musculus*) stranded in the St. Lawrence estuary, Quebec," *J Helminthol Soc Wash*, vol. 59, no. 2, pp. 206-211.

_____, 1993, "Annotated list of Metazoan parasites reported from the Blue whale, *Balaenoptera musculus*," *J Helminthol Soc Wash*, vol. 60, no. 1, pp. 62-66.

MEASURES, L.N., P. BÉLAND, D. MARTINEAU and S. DE GUISE, 1995, "Helminths of an endangered population of belugas, *Delphinapterus leucas*, in the St. Lawrence estuary, Canada," *Can J Zool*, 73, pp. 1402-1409.

MICHALEV, J.A., 1979, "Revealing of differences in the Antarctic baleen whale stock on the basis of the analysis of the Jacobson's organ position," *Report of the International Whaling Commission*, vol. 29, pp. 345-346.

MICHAUD, R., 1993, *Rencontre avec les baleines du Saint-Laurent*, Tadoussac (QC), GREMM, 75 p.

MILINKOVITCH, M. 1995, "Molecular phylogeny of cetaceans prompts revision of morphological transformations," *TREE*, vol. 10, no. 8 (August), pp. 328-334.

_____, 1996, "Baleines, cachalots et dauphins. La biologie cellulaire révèle d'étonnantes parentés entre cétacés très différents," *La Recherche* 288 (June), pp. 42-43.

MILINKOVITCH, M., C. GUILLERMO ORTI and A. MEYER, 1993, "Revised phylogeny of whales suggested by mitochondrial ribosomal DNA sequences," Letters to Nature, *Nature*, vol. 361 (January 28), pp. 346-348.

MILINKOVITCH, M., A. MEYER and J.R. POWELL, 1994, "Phylogeny of all major groups of cetaceans based on DNA sequences from three mitochondrial genes," *Mol Biol Evol* 11, pp. 939-948.

MILINKOVITCH, M., J.L. DUNN and J.R. POWELL, 1994, "Exfoliated cells as the most accessible DNA source for captive whales and dolphins," *Marine Mammal Science*, vol. 10, no. 1 (January) pp. 125-128.

MILINKOVITCH M., G. ORTI and A. MEYER, 1995, "Novel phylogeny of whales revisited but not revised," *Mol Biol Evol*, vol. 12, no. 3, pp. 518-520.

MILLER, G.S. Jr., 1923, *The Telescoping of the Cetacean Skull*, Washington, Smithsonian Institution, Publication 2720.

MIZROCH, S., 1981, "Analysis of some biological parameters of the Antarctic fin whale (*Balaenoptera physalus*):" *Reports of the I.W.C.* 31, pp. 425-434.

NAKAMURA, T., 1984, *Gentle Giant at Sea with the Humpback Whale*, San Francisco, Chronicle Books, 66 p.

NISHIWAKI, M., T. SCHIBARA and S. OSHUMI, 1958, "Age studies of fin whales based on ear plugs," *Scientific Reports of the Whale Research Institute* 13, pp. 155-169.

NORTHRIDGE, S. and G. PILLERI, 1986, "A review of human impact on small cetaceans," in G. PILLERI (dir.), *Investigations on Cetacea*, vol. XVIII, Berne.

OHLAND, D.P., E. HARLEY and P.B. BEST, 1995, "Systematics of cetaceans using restriction site mapping of mitochondrial DNA molecular phylogenetics and evolution," *Mol Biol Evol*, vol. 4, no. 1 (March), pp. 10-19.

OSHUMI, S. 1964, "Examination of age determination of the fin whale," *Scientific Reports of the Whale Research Institute* 18, pp. 49-88.

PILLERI, G., K. ZBINDEN and C. KRAUS, 1979, "The sonar field of *Inia geoffrensis*," *Investigations on Cetacea*, vol. X, pp. 157-176.

PILLERI, G., K. ZBINDEN and H. MINGLONG, 1983, "The sonar field in the bottlenose dolphin *Tursiops truncatus*," *Investigations on Cetacea*, vol. XV, pp. 81-94.

PILLERI, G., K. ZBINDEN, M. GIHR and C. KRAUS, 1976, "Sonar cliks, directionality of the emission field and echolocating behaviour of the Indus dolphin (*Platanista indi*, Blyth, 1859)," *Investigations on Cetacea*, vol. VII, pp. 13-43.

PROULX, J.-P., 1986, *Whaling in the North Atlantic from the Earliest Times to the Mid-19th Century*, Environment Canada, Parks Canada, Studies in Archeology, Architecture and History, National Historic Parks and Sites Branch. 119 p.

PURVES, P.E. and G.E. PILLERI, 1983, *Echolocation in Whales and Dolphins*, Academic Press, 258 p.

QUAY, W.B. and E.D. MITCHELL, 1971, "Structure and sensory apparatus of oral remnants of the nasopalatine canals in the fin whale (*Balaenoptera physalus* L.)," *Journal of Morphology* 134, pp. 271-280.

REIDENBERG J.S. and J.T. LAITMAN, 1987, "Position of the larynx in Odontoceti (toothed whales)," *The Anatomical Record* 218, pp. 98-106.

_____, 1988, "Existence of vocal folds in the larynx of Odontoceti (toothed whales)," *The Anatomical Record* 221, pp. 884-891.

_____, 1994, "Anatomy of the hyoid apparatus in Odontoceti (toothed whales). Specialisation of their skeleton and musculature," *The Anatomical Record* 240, pp. 598-624.

RIDGWAY, S.H., 1972, "Homeostasis in the aquatic environment," in S.H. RIDGWAY (dir.), *Mammals of the Sea: Biology and Medicine*, Springfield (IL), C.C. Thomas, pp. 590-747.

_____, 1972, *Mammals of the Sea: Biology and Medicine*, Springfield (IL), C.C. Thomas.

ROBERTSON, R.B., 1955, *Avec les chasseurs de baleines*, Paris, Amyot Dumont, 208 p.

ROMER, A.S., 1959, *The Vertebrate Story*, London, University of Chicago Press, 437 p.

ROSS, W.G., 1985, *Arctic Whalers, Icy Seas*, Toronto, Irwin Publishing, 262 p.

SIMARD, Y., 1994, "Comment la mer nourrit-elle les baleines à Tadoussac ou le pourquoi océanographique de la visite estivale des rorquals dans l'estuaire maritime du Saint-Laurent, à la tête du chenal laurentien (Tadoussac, Les Escoumins, Grandes-Bergeronnes)," *L'Euskarien*, Société Provancher d'histoire naturelle (Summer) pp. 33-38.

SIMARD Y., R. DE LADURANTAYE and J.-C. THÉRIAULT, 1986, "Aggregation of euphausiids along a coastaal shelf in an upwelling environment," *Marine Ecology-Progress Series*, vol. 32, pp. 203-215.

SIMARD, Y., G. LACROIX and L. LEGENDRE, 1986, "Diel vertical migration and nocturnal feeding of a dense coastal krill scattering layer (*Thysanoessa raschi* and *Meganyctiphanes norvegica*) in stratified surface waters," *Marine Biology* 91, pp. 93-105.

SLIJPER, E.J., 1978, *Whales and Dolphins*, Ann Arbor, The University of Michigan Press, 170 p.

STOOPS, E.D., J.L. MARTIN and D.L. STONE, 1996, *Whales*, Sterling Publishing Co., 80 p.

SUMICH, J.L., 1980, *An Introduction to the Biology of Marine Life*, W.C. Brown Company Publishers, 357 p.

SYLVESTRE, J.-P., 1989, *Baleines et cachalots*, Lausanne, Delachaux et Niestlé, 135 p.

THEWISSEN, J.G.M., L.J. ROE, J.R. O'NEIL, S.T. HUSSAIN, A. SAHNI and S. BAJPAI, 1996, "Evolution of cetacean osmoregulation," *Nature*, vol. 381, pp. 379-380.

TINKER, S.W., 1988, *Whales of the World*, Bess Press, 310 p.

TOMILINE, A., 1977, *Le monde des baleines et des dauphins*, Éditions Mir Moscou, 287 p.

TRUE, F.W., 1983, *The Whalebone Whales of Western*, Washington (DC), The North Atlantic Smithsonian Institution Press, 332 p.

TUCK, J.A., and R. GRENIER, 1989, *Red Bay, Labrador, World Whaling Capital A.D. 1550-1600*, St. John's (NF.), Atlantic Archaeology Ltd., 68 p.

VAUGHAN, T.A., 1986, *Mammalogy*, New York, Saunders College Publishing, 576 p.

VLADYKOV, V. D., 1944, *Chasse, biologie et valeur économique du marsouin blanc ou béluga (*Delphinapterus leucas*) du fleuve et du golfe Saint-Laurent*, Province de Quebec, Department of Fisheries of Québec.

_____, 1946, *Nourriture du marsouin blanc ou béluga (*Delphinapterus leucas*) du fleuve Saint-Laurent*, Department of Fisheries of Québec, 132 p.

WARD, N., 1995, *Stellwagen Bank*, Provincetown (MA), Down East Books, Center for Coastal Studies, 232 p.

WATSON L., 1981, *Sea Guide to the Whales of the World*, Nelson Canada Limited, 304 p.

WEBB, R.L., 1988, *On the Northwest, Commercial Whaling in the Pacific Northwest, 1790-1967*, Vancouver, University of British Columbia Press, 425 p.

WILLIAMS, H., 1988, *Des baleines*, Aubier, 191 p.

ZBINDEN K., G. PILLERI and C. KRAUS, 1980, "The sonar field in the white whale (*Delphinapterus leucas*) (Pallas, 1776)," *Investigations on Cetacea*, vol. XI, pp. 124-155.

ZOEGER, J., 1981, "Magnetic material in the head of the common Pacific dolphin (*Delphinus delphis*)," *Science*, vol. 213, pp. 892-894.

Suggested Web Sites on the Internet

WHALENET
  http://whale.wheelock.edu
MARINE ENVIRONMENT RESEARCH AND EDUCATION CENTER (ANITAK) (Spain)
  http://www.lander.es:800/~canadas/ingles.html
INTERDISCIPLINARY CENTER FOR BIOACOUSTICS AND ENVIRONMENTAL RESEARCH (Italy)
  http://cibra.unipv.it

COASTAL ECOSYSTEMS RESEARCH FONDATION
http://www.cerf.bc.ca
EARTH WATCH INSTITUTE
http://www.earthwatch.org/t/Tfieldsofstudy.html
WHALE RESEARCH GROUP(NEWFOUNDLAND)
http://play.psych.mun.ca/Psych/whale.html
SOCIETY FOR MARINE MAMMALOGY
http://pegasus.cc.ucf.edu/~smm/const.htm
WHALES ON THE NET-WORLD NEWS ARCHIVES
http://whales.magna.com.au/news/index.html
HIGH NORTH WEB
http://www.highnorth.no/hnn11.htm
WHALE CLUB
http://www.whaleclub.com/index.html
WHALE CONSERVATION INSTITUTE OF ROGER PAYNE
http://www.whale.org
AMERICAN CETACEAN SOCIETY
http://www.acsonline.org/
EUROPEAN CETACEAN SOCIETY
http://web.inter.NL.net/users/J.W.Broekema/ecs.htm
INTERNATIONAL WHALING COMMISSION
http://ourworld.compuserve.com/homepages/iwcoffice/
WHALE WATCHING WEB
http://www.physics.helsinki.fi/whale/
MARMAM DISCUSSION LIST, ARCHIVES
http://members.aol.com/marmamnews/index.html
GROUPE DE RECHERCHE ET D'ÉDUCATION SUR LE MILIEU MARIN
gremm@fjord-best.com

MINGAN ISLANDS RESEARCH SOCIETY
378, bord de la Mer
Longue-Pointe-de-Mingan (Québec)
Canada G0G 1V0
Tel: (418) 949-2845
Fax: (418) 949-2845

For all questions contact
Mingan Islands Research Society
285, rue Green
St-Lambert (Québec)
Canada J4P 1T3
Tel: (450) 465-9176
Fax: (450) 465-9176
email: Richard Sears
rsblues@polysoft.com